A HANDBOOK OF MODERN IMMUNOLOGY

A HANDBOOK OF MODERN IMMUNOLOGY

Bijender Bahl

CYBER TECH PUBLICATIONS
4264/3, Ansari Road, Daryaganj, New Delhi-110002 (India)
Ph.: 011-23244078, 011-43559448 Fax: 011-23280028
E-Mail: cyberpublicationsdelhi@yahoo.com
Website: www.cybertechpublications.com

A HANDBOOK OF MODERN IMMUNOLOGY

Bijender Bahl

First Edition 2010

Published by :

G.S. Rawat for **Cyber Tech Publications**
4264/3, Ansari Road, Daryaganj, New Delhi-110002 (India)
Ph.: 011-23244078, 011-43559448 Fax: 011-23280028
E-Mail: cyberpublicationsdelhi@yahoo.com
Website: www.cybertechpublications.com

Preface

Immunological science is in the midst of a revolution. The introduction of monoclonal antibody methodology, the application of the techniques of molecular and somatic cell genetics, and the development of methods to grow cloned lines of lymphocytes in long-term culture have made possible experiments that, even a few years ago, would have been regarded as science fiction. Together with the application of modern biochemical technology to the study of immunologic phenomena, these approaches are reshaping our understanding of the immune system. The advances achieved by immunologists are of enormous importance in understanding the biological basis of this most fascinating system. Equally important are their impact in the clinical sphere. The immune system is the most important defense against infections by pathogenic microorganisms, may be critical in prevention and therapy of neoplasms, and is intimately involved in the pathogenesis of a wide range of chronic diseases. Its primacy in these areas is tragically emphasized by the recent emergence of a new human disorder—acquired immunodeficiency syndrome.

The enormous expansion of immunology as both a basic biological discipline and as one of the most important of the clinical sciences makes it virtually impossible for even the most diligent immunologist to remain abreast of important developments outside his or her area of special interest. The problem of keeping up to date with this critical science is even greater for individuals in allied fields and for students.

The principal areas of immunology by active and respected scientists so that major topics are covered on a periodic basis. The humanistic aspects of our field and for a presentation of the major developments of immunology as they happened.

—Author

CONTENTS

1

Immunology

EVOLUTION OF THE IMMUNE SYSTEM

The immune system consists of factors that provide innate and acquired immunity, and has evolved to become more specific, complex, efficient, and regulated. One of the principal functions of the human immune system is to defend against infecting and other foreign agents by distinguishing self from non-self (foreign antigens) and to marshal other protective responses from leukocytes. The immune system, if dysregulated, can react to self antigens to cause autoimmune diseases or fail to defend against infections.

Organization/Components/Functions

The immune system is organized into discrete compartments to provide the milieu for the development and maintenance of effective immunity. Those two overlapping compartments: the lymphoid and reticuloendothelial systems (RES) house the principal immunologic cells, the leukocytes. Leukocytes derived from pluripotent stem cells in the bone marrow during postnatal life include neutrophils, eosinophils, basophils, monocytes and macrophages, natural killer (NK) cells, and T and B lymphocytes. Hematopoietic and lymphoid precursor cells are derived from pluripotent stem cells. Cells that are specifically committed to each type of leukocyte (colony-forming units) are consequently produced with the assistance of special stimulating factors (e.g. cytokines).

Cells of the immune system intercommunicate by ligand-receptor interactions between cells and/or via secreted molecules called cytokines. Cytokines produced by lymphocytes are termed lymphokines (i.e., interleukins and interferon-gamma) and those produced by monocytes and macrophages are termed monokines.

Lymphoid System

Cells of the lymphoid system provide highly specific protection against foreign agents and also orchestrate the functions of other parts of the immune system by producing immunoregulatory cytokines. The lymphoid system is divided into 1) central lymphoid organs, the thymus and bone marrow, and 2) peripheral lymphoid organs, lymph nodes, the spleen, and mucosal and submucosal tissues of the alimentary and respiratory tracts. The thymus instructs certain lymphocytes to differentiate into thymus-dependent (T) lymphocytes and selects most of them to die in the thymus (negative selection) and others to exit into the circulation (positive selection). T lymphocytes circulate through the blood, regulate antibody and cellular immunity and help defend against many types of infections. The other classes of lymphocytes, B cells (antibody-forming cells) and natural killer (NK) cells, are thymic-independent and remain principally in peripheral lymphoid organs.

RETICULOENDOTHELIAL SYSTEM

Cells of the RES provide natural immunity against microorganisms by (1) a coupled process of phagocytosis and intracellular killing, (2) recruiting other inflammatory cells through the production of cytokines, and (3) presenting peptide antigens to lymphocytes for the production of antigen-specific immunity. The RES consists of (1) circulating monocytes; (2) resident macrophages in the liver, spleen, lymph nodes, thymus, submucosal tissues of the respiratory and alimentary tracts, bone marrow, and connective tissues; and (3) macrophage-like cells including dendritic cells in lymph nodes, Langerhans cells in skin, and glial cells in the central nervous system.

Leukocytes

Leukocytes, the main cells in the immune system, provide either innate or specific adaptive immunity. These cells are derived from myeloid or lymphoid lineage. Myeloid cells include highly phagocytic, motile neutrophils, monocytes, and macrophages that provide a first line of defense against most pathogens. The other myeloid cells, including eosinophils, basophils, and their tissue counterparts, mast cells, are involved in defense against parasites and in the genesis of allergic reactions. In contrast, lymphocytes regulate the action of other leukocytes and generate specific immune responses that prevent chronic or recurrent infections.

Myeloid Cells

Neutrophils: These are one of the major types of cells that are recruited to ingest, kill, and digest pathogens. Neutrophils are the most highly adherent, motile, phagocytic leukocytes and are the first cells recruited to acute inflammatory sites. Each of their functions is dependent upon special proteins, such as the adherence molecule CD11b/CD18, or biochemical pathways, such as the respiratory burst associated with cytochrome b_{558}.

Eosinophils: Eosinophils defend against many types of parasites and participate in common hypersensitivity reactions via cytotoxicity. That cytotoxicity is mediated by large cytoplasmic granules, which contain the eosinophilic basic and cationic proteins.

Basophils: These cells and their tissue counterparts, mast cells, produce cytokines that help defend against parasites and engender allergic inflammation. These cells display high affinity surface membrane receptors for IgE antibodies and have many large cytoplasmic granules, which contain heparin and histamine. When cell-bound IgE antibodies are cross-linked by antigens, the cells degranulate and produce low-molecular weight vasoactive mediators (e.g. histamine) through which they exert their biological effects.

Monocytes/Macrophages: Monocytes and macrophages are involved in phagocytosis and intracellular killing of microorganisms. Macrophages process protein antigens and present peptides to T cells. These monocytes/macrophages are highly adherent, motile and phagocytic; they marshal and regulate other cells of the immune system, such as T lymphocytes; serve as antigen processing-presenting cells; and act as cytotoxic cells when armed with specific IgG antibodies.

Macrophages are differentiated monocytes, which are one of the principal cells found to reside for long periods in the RES. Macrophages may also be recruited to inflammatory sites, and be further activated by exposure to certain cytokines to become more effective in their biologic functions.

Lymphoid Cells

These cells provide efficient, specific and long-lasting immunity against microbes and are responsible for acquired immunity. Lymphocytes differentiate into three separate lines: thymic-dependent cells or T lymphocytes that operate in cellular and humoral immunity, B lymphocytes that differentiate into plasma cells to secrete antibodies, and natural killer (NK) cells. T and B lymphocytes are the only lymphoid cells that produce and express specific receptors for antigens.

T Lymphocytes: These cells are involved in the regulation of the immune response and in cell mediated immunity and help B cells to produce antibody (humoral immunity). Mature T cells express antigen-specific T cell receptors (TcR) that are clonally segregated (i.e., one cell lineage-one receptor specificity). Every mature T cell also expresses the CD3 molecule, which is associated with the TcR. In addition mature T cells display one of two accessory molecules, CD4 or CD8. The TcR/CD3 complex recognizes antigens associated with the major histocompatibility complex (MHC) molecules on target cells (e.g. virus-infected cell). The TcR is a transmembrane heterodimer composed of two polypeptide chains (usually, alpha and beta chains). Each chain consists of a constant (C) and a variable (V) region, and are formed by a gene- sorting mechanism similar to that found in antibody formation. The repertoire is generated by combinatorial joining of variable (V), joining (J), and diversity (D) genes, and by N region (nucleotides inserted by the enzyme deoxynucleotidyl-transferase) diversification. Unlike immunoglobulin genes, genes encoding TcR do not undergo somatic mutation. Thus there is no change in the affinity of the TcR during activation, differentiation, and expansion.

T Helper Cells: These cells are the primary regulators of T cell- and B cell-mediated responses. They 1) aid antigen-stimulated subsets of B lymphocytes to proliferate and differentiate toward antibody-producing cells; 2) express the CD4 molecule; 3) recognize foreign antigen complexed with MHC class II molecules on B cells, macrophages or other antigen-presenting cells; and 4) aid effector T lymphocytes in cell-mediated immunity. Currently, it is believed that there are two functional subsets of T helper (Th) cells. Th1 cells aid in the regulation of cellular immunity, and Th2 cells aid B cells to produce certain classes of antibodies (e.g., IgA and IgE). The functions of these subsets of Th cells depend upon the specific types of cytokines that are generated, for example interleukin-2 (IL-2) and interferon-gamma (IFN-gamma) by Th1 cells and IL-4 and IL-10 by Th2 cells.

Cell-mediated immunity (delayed hypersensitivity) plays an important role in defense against many intracellular infections such as Mycobacterium tuberculosis. This inflammatory reaction is initiated by the recognition of specific antigens by Th1 cells. Consequently, lymphokines are generated which recruit activated macrophages to eliminate foreign antigens or altered host cells.

T Cytotoxic Cells: These cells are cytotoxic against tumor cells and host cells infected with intracellular pathogens. These cells 1) usually express CD8, 2) destroy infected cells in an antigen-specific manner that is dependent upon the expression of MHC class I molecules.

T Suppressor Cells: These cells suppress the T and B cell responses and express CD8 molecules.

Natural Killer Cells: NK cells are large granular lymphocytes that nonspecifically kill certain types of tumor cells and virus-infected cells. Killing by NK cells is enhanced by cytokines such as IL-2 and IFN-gamma. NK cells are also activated by microorganisms to produce a number of cytokines [(IL-2, IFN-gamma, IFN-alpha, and tumor necrosis factor-alpha

(TNF-alpha)]. These circulating large granular lymphocytes do not express CD3, TcR or immunoglobulin, but display surface receptors (CD16) for the Fc fragment of IgG antibodies.

B Lymphocytes: These cells differentiate into plasma cells to secrete antibodies and are involved in processing proteins and presenting the resultant peptide antigen fragments in the context of MHC molecule to T cells. The genesis of μ and delta chain-positive, mature B cells from pre-B cells is antigen-independent. Pre-B cells in the bone marrow undergo gene rearrangement for IgM heavy (H) chains and consequently express those proteins in the cytoplasm (the μ chain), but no immunoglobulin light (L) chains. B cell development is characterized by recombinations of immunoglobulin H and L chain genes and expression of specific surface monomeric IgM molecules. At this stage of development, B cells are highly susceptible to the induction of tolerance. Once these cells acquire IgD molecules on their surface, they become mature B cells that are able to differentiate after exposure to antigen into antibody-producing plasma cells.

The activation of B cells into antibody producing/secreting cells (plasma cells) is antigen-dependent. Once specific antigen binds to surface Ig molecule, the B cells differentiate into plasma cells that produce and secrete antibodies of the same antigen-binding specificity. If B cells also interact with Th cells, they proliferate and switch the isotype (class) of immunoglobulin that is produced, while retaining the same antigen-binding specificity. This occurs as a result of recombination of the same Ig VDJ genes (the variable region of the Ig) with a different constant (C) region gene such as IgG. In the case of protein antigens, Th2 cells are thought to be required for switching from IgM to IgG, IgA, or IgE isotypes.

IgM is therefore the principal antibody produced during a primary immunization. This primary antibody response is manifested by serum IgM antibodies as early as 3-5 days after the first exposure to an immunogen (immunizing antigen), peaks in 10 days, and persists for some weeks. Secondary or anamnestic antibody responses following repeated exposures to the same antigen appear more rapidly, are of longer duration, have higher affinity, and principally are IgG molecules.

When antibodies bind to antigens, they may (1) neutralize pathogenic features of antigens such as their toxins, (2) facilitate their ingestion by phagocytic cells (opsonization), (3) fix to and activate complement molecules to produce opsonins and chemoattractants (vide infra), or (4) participate in antibody-dependent cellular cytotoxicity (ADCC).

In addition to antibody formation, B cells also process and present protein antigens. After the antigen is internalized it is digested into fragments, some of which are complexed with MHC class II molecules and then presented on the cell surface to $CD4^+$ T cells.

IMMUNOGLOBULIN SUPERGENE FAMILY

Immunoglobulins (Ig)/Antibodies

Immunoglobulins (antibodies) are globular glycoproteins found in body fluids or on B cells where they act as antigen receptors. These molecules are either expressed on the surface of B cells or are secreted by terminally differentiated cells from this lineage (plasma cells) into the circulation or external secretions. An immunoglobulin molecule is a symmetrical multi-chain peptide consisting of two identical H chains and two identical L chains. Each chain is divided into a V region that is responsible for specific antigen binding and a C region that carries out other functions such as the binding of IgG to complement or leukocytes. These antibody molecules are formed as a result of the assembly of separate germ-line genes for the V, J, and C regions of the H and L chains of the final immunoglobulin molecule. This combinatorial mechanism is responsible for the great diversity of antibody molecules.

There are five major isotypes (classes) of immunoglobulins (IgG, IgA, IgM, IgD, and IgE). These isotypes are distinguished by differences in the C regions of H chains of each immunoglobulin isotype (gamma, alpha, μ, delta, and epsilon, respectively). These differences are responsible for the particular functions of immunoglobulin classes.

T Cell Receptor

The specific receptor for antigen on T lymphocytes, the TcR, is a heterodimeric protein with motifs that are similar to immunoglobulin molecules, but whose structure is encoded by a different set of V, J, D, and C genes. Moreover, T cells consist of two subsets carrying different receptors, that have been designated alpha/beta and gamma/delta. The T cell receptors act as specific antigen recognition molecules. Unlike antibody molecules, the TcR molecules cannot recognize soluble antigens. In contrast, they recognize protein antigens that have been processed and presented as peptides on the surface of antigen-presenting cells in the context of MHC class I or MHC class II molecules (vide infra).

Major Histocompatibility Complex (MHC)

These genes encode for cell surface molecules that are involved in the genesis and regulation of specific immune responses to T-cell dependent antigens and in tissue transplantation. They principally encode cell surface protein molecules that bind antigenic peptides, which are recognized by T cells.

The MHC is a cluster of ~ 40-50 genes located on chromosome 6. These genes belong to the super-immunoglobulin gene family. There are three classes of these molecules. MHC class I molecules are found on all nucleated somatic cells and aid in presenting endogenously synthesized antigens, whereas MHC class II molecules are found principally on antigen processing/presenting cells (i.e., macrophages, B cells) and are involved in presenting processed exogenous protein antigens. The MHC class III region contains a heterogeneous group of genes that encode for some components of the complement system, heat shock proteins, tumor necrosis factor-alpha, and tumor necrosis factor-beta.

T Cell Activation

The presentation of antigen in the context of MHC molecules is essential for T cell recognition of peptide antigens. However, interactions between the MHC-bound peptide and

TcR and the MHC class I or class II molecules with CD8 or CD4, respectively, is not sufficient to activate T cells. Other ligands on antigen presenting cells and their receptors (co-receptors/co-stimulators) on T cells are required to complete the process of T cell activation.

Tolerance-Autoimmunity

Immunologic tolerance (unresponsiveness) normally prevents reactions against self-antigens; if immunologic tolerance is broken, autoimmune reactions may occur. Much of the development of tolerance occurs in the thymus by the elimination (clonal deletion) or inactivation (clonal anergy) of self-reactive clones of T cells. Other mechanisms of tolerance occur extrathymically and include activation of antigen-specific T suppressor cells and clonal deletion, which results in the elimination of self-reactive B cells or T cells, and clonal anergy.

Tolerance may be broken because of a genetic predisposition to immune dysregulation, altered self-antigens, exposure to microbial antigens that cross-react with self-antigens, or exposure to a self-antigen that is normally not revealed to the immune system (e.g., an antigen in the eye). When tolerance against self-antigens is broken, autoimmunity is produced, which could result in an autoimmune disease.

The Complement System

The complement system consists of inactive circulating glycoproteins that can be sequentially activated by antigen-antibody (IgG or IgM) complexes or bacterial products to enhance inflammation or to attack cellular membranes. The system consists of the classical and alternative pathways that converge to activate the membrane attack complex. After activation, opsonic, chemoattractant, or cytotoxic fragments are produced.

Defenses Against Infections

Natural (innate) and acquired defenses are marshalled to combat infecting agents. The first line of defense includes the skin, mucous membranes, protective inhibitors, and IgA antibodies produced at mucosal sites. The second line of defense consists of local factors and cells that are activated or recruited to the site of microbial invasion. These include: (1) the coagulation system, (2) the fibrinolytic system, (3) vasoactive peptides, (4) the complement system, (5) resident macrophages, (6) recruited inflammatory leukocytes, and (7) cytokines. The third line of defense includes the expansion of populations of antigen-specific B cells and T cells, the production of systemic antibodies, and the activation of T cells. Successful defense is followed by a clearance of opsonized pathogens by the RES and tissue repair.

IMMUNE RESPONSES TO MICROORGANISMS LEAD TO DISEASE

Excessive or otherwise inappropriate immune responses to infecting agents may lead to disease. Examples of such excessive immunologic responses that can be protective or cause disease include:

(1) circulating antigen-antibody (immune) complexes of microbial antigens bound to IgM or IgG antibodies,
(2) antibodies to microorganisms that cross-react with self-antigens,
(3) vasoactive compounds from the complement system and from the metabolism of arachidonic acid,
(4) excessive production of proinflammatory cytokines,
(5) delayed hypersensitivity reactions, and
(6) cytotoxic T cells directed against the infected host cells.

Ontogeny of the Immune Response

The immune system undergoes an orderly development during the prenatal and postnatal periods. Mature T and B cells appear in the fetus, but are not activated until the infant is exposed to immunogens. Memory T cells are not present during early infancy and the antibody repertoire is not fully established for many months. IgM is the first type of antibody produced postnatally. IgG antibodies to protein antigens are formed in early infancy, but IgG antibodies to polysaccharides do not appear until 2 - 2.5 years of age. There are also developmental delays in the production of several cytokines such as the interferons.

Maternal Immunologic Contributions to the Infant

Maternal immune factors are transmitted to the fetus via the placenta and to the young infant by mammary gland secretions. These transferred maternal factors compensate for developmental delays in the production of those immune factors by the recipient fetus/infant. Developmental delay in the production of IgG is overcome by transfer of maternal antibodies of that same isotype via the placenta. Other immune factors (whose production is developmentally delayed), such as secretory IgA, lactoferrin, and lysozyme; leukocytes; anti-inflammatory agents; and immunomodulating agents are provided by mammary gland secretions via human milk. These factors are not as well represented in non-human milk. Therefore, the breast-fed infant is less at risk for gastrointestinal and respiratory infections and for inflammatory disorders including common allergic diseases.

Immunologic Deficiency

Immune deficiencies are genetic or acquired and result in an increased susceptibility to certain infections, the types of which depend upon the exact nature of the defect.

Genetic Defects: X-linked agammaglobulinemia is a genetic defect in a B cell progenitor kinase that is essential for B cell development. Consequently, few B cells and only low levels of antibodies are produced. This leads mainly to an increased susceptibility to highly virulent, encapsulated respiratory bacterial infections. T cell deficiency is the primary problem in severe combined immunodeficiency (deficiencies of B and T cells). Most cases are due to an X-linked recessive defect in the formation of the gamma-chain common to a number of cytokine receptors. Some autosomal recessive types are due to deficiencies in enzymes such as adenosine deaminases in the purine salvage pathway. Patients with these diseases display few T cells, decreased T cell functions, poor antibody formation, and an increased susceptibility to opportunistic infections such as Pneumocystis carinii.

Hereditary defects also occur in neutrophils. For example, a decrease in leukocyte adherence is due to an autosomal recessive defect in the formation of the common CD18-subunit of leukocyte adherence glycoproteins, whereas deficiency in intracellular killing (chronic granulomatous disease) is due to a deficiency in the production of subunits of cytochrome b_{558} or ancillary proteins necessary for their stabilization. Consequently, reactive oxygen compounds required for intracellular killing are not produced.

Acquired Defects: Protein-energy malnutrition is the leading cause of immunologic deficiency. A second, but important cause of acquired immunodeficiency is the human immunodeficiency virus (HIV) that attacks $CD4^+$ T cells and macrophages. Also, certain other infections depress or destroy parts of the immune system by different mechanisms.

EVOLUTION OF THE IMMUNE SYSTEM

The human defense system consists of factors that provide innate and acquired immunity against microorganisms. The system evolved from primitive but effective defenses found in more ancient animal species. The innate defenses include (1) structural barriers, (2) acids, bases, and other chemical agents produced at various sites, such as mucosal surfaces, and (3) highly phagocytic, motile scavenger cells that have well-developed killing and digestive powers. As a result of the evolutionary process, the mammalian immune system has become more specific, efficient, regulated, and complex. The development of specialized innate and acquired recognition/ regulatory proteins (antibodies, cell receptors, and cytokines) expanded the repertoire, and control the magnitude of the protective responses. One of the most important consequences of this evolution is the ability of the immune system to discriminate between self and non-self antigens and maintain a memory of previous encounters with antigens, including those from microorganisms.

The evolutionary changes allowed development of B and T cells which express antigen-specific receptors on their cell surface. These changes permit humans to survive in an environment laden with microbial pathogens and environmental toxins. The pathogenic features of those microorganisms include the ability to (1) enter the body through portals such as the skin, respiratory system, and the alimentary tract; (2) utilize nutrients from those sites; (3) adhere to epithelium; (4) produce virulence factors and toxins; (5) commandeer the replicative machinery of the host's cells; (6) evade the immunologic system; (7) cripple the defenses of the host; and (8) cause autoimmune responses by acting as cross-reactive antigens.

The salient features of the human defense system that evolved to counteract the pathogenic microorganisms and prevent autoimmune problems will be presented in the rest of this chapter.

Organization/Components/Functions

The production, maturation, and function of cells of the immune system occur to a great extent in two overlapping organ systems, (1) the lymphoid system consisting of lymphocytes and their supporting structures and 2) the RES consisting of macrophages and related mononuclear phagocytes. In postnatal life, bone marrow is the principal source of pluripotent stem cells that produce precursors of cells that operate in host defense. The development of each type of leukocyte is precisely controlled and the controls account for the great specificity of the defense system and the fact that untoward immunologic reactions are relatively uncommon.

Lymphoid System

The lymphoid system consists of organs that house (1) T and B cells that are responsible for antigen-specific immunity and (2) NK cells that are innately cytotoxic to tumor cells and cells expressing foreign antigens. The system is divided into (a) central lymphoid organs, the thymus and bone marrow, and (b) peripheral lymphoid organs including lymph nodes, spleen, and the mucosa/submucosa of the respiratory and alimentary tracts. Lymphocytes are one of the principal leukocytes found in these organs. There are three major types of lymphocytes (T, B, and NK) that have distinctive surface markers and functions (see sections on T cells, B cells, and NK cells. Furthermore, the T and B cells in the lymph nodes are confined to discrete zones.

RETICULOENDOTHELIAL SYSTEM (RES)

The second major cellular system, the RES, harbors macrophages, which are cells that play major roles in (1) defending against many microbial pathogens and (2) generating specific immune responses by processing protein antigens and presenting the resultant peptide antigen fragments in the context of MHC molecules to T cells. The system consists of (1) monocytes in the blood, (2) macrophages in the liver, spleen, lymph nodes, thymus, bone marrow, connective tissues, and submucosal tissues of the respiratory and alimentary tracts, (3) dendritic cells in lymph nodes, (4) Langerhans cells in skin, and (5) glial cells in the central nervous system. Macrophages not only operate in direct defense (phagocytosis and intracellular killing) but also marshal other parts of the immune system, such as T lymphocytes.

Molecular Communications in the Immune System

Cells of the immune system have profound immunoregulatory influences on each other. This regulation is mediated in large part by potent bioactive molecules, called cytokines, which may have autocrine, paracrine, or systemic effects. These polypeptides and glycoproteins are produced by diverse types of cells and act on many different types of cells by binding to high affinity receptors on their surfaces. Their functions include (1) activating or attracting specific types of leukocytes, (2) regulating cell division, (3) modulating the production or actions of other cytokines, (4) promoting or abrogating inflammation, (5) directing certain cells to switch the types of proteins that they produce, and (6) influencing the production of cellular or humoral immunity. Those produced principally by lymphocytes have been termed lymphokines and those principally produced by monocytes and macrophages, monokines. Interleukin is also used to designate many of these agents.

Ligands, such as surface molecules, on certain cells of the immune system that bind to receptors on other types of cells may activate the cells bearing the receptors and thus modulate immune responses. Interactions between T cell receptors and the processed peptide in the context of the MHC molecule are examples of the importance of such ligand (peptide)-receptor (T cell receptor) interactions in the immune system.

Immunoglobulin Supergene Family

The immunoglobulin supergene family is a group of structurally similar glycoproteins, which mediate antigen recognition and cellular interactions. They are derived from a family of genes which evolved from a common primordial gene. The products of these genes are transmembrane glycoproteins characterized by a common structural motif of functional domains. Some important members of this immunoglobulin supergene family are the immunoglobulins, MHC, TcR, secretory component, and adherence proteins such as ICAM-1.

IMMUNOGLOBULINS

Structures

The basic structure of all immunoglobulin molecules consists of two identical L chains and two identical H chains. The antibody molecule consists of three major domains connected by a hinge region. As shown in Fig. 1-15, digestion of antibody molecule with a proteolytic enzyme-papain results in the separation of these three domains. Two domains are identical and are called fragment antigen binding (Fab), and the third domain is called fraction crystallizable (Fc). However, treatment with proteolytic enzyme pepsin results in a fragment that contains both antigen binding arms (Fab')2 and several pieces of the Fc fragment. Fab interact with the antigen, and Fc bind to Fc-receptors on different cells.

Various forms of immunoglobulins such as IgG and IgE are found as monomers, secreted IgA as dimers and IgM as pentamers. Consequently, two distinct regions of the assembled immunoglobulin occur: the first, which binds to an antigenic determinant and the second, which has other functions, such as binding to special cells and the first component of complement. The two H chains and each H chain and L chain are linked by disulfide bonds. Each chain is divided into two regions: the C region at the carboxyl-terminus and the variable region at the amino-terminus. The C region of each L chain consists of about 107 amino acids and has an invariant structure except for isotypic features (kappa or lambda) and allotypic variants (e.g., molecular structures that are individually inherited). V and C regions of H chains are further divided into domains characterized by folding of the polypeptide chain into 110 amino acid loops. V regions of H and L chains display great variability in the sequence of amino acids. Localized areas of these hypervariable regions of H and L chains interact to form antigen binding sites. In contrast, C regions of H chains dictate other functions of immunoglobulins, including binding to cell surface receptors. Eight immunoglobulin isotypes, IgG1, IgG2, IgG3, IgG4, IgA1, IgA2, IgM, IgD, and IgE, are produced by B cells as a result of rearrangements of V genes for H chains (VH), D genes for H chains (DH), J genes for H chains (JH), V genes for L chains (VL), J genes for L chains (JL), and C region genes (vide infra) The special properties of each immunoglobulin class are as follows.

IgG: IgG is a monomeric, four-chain structure consisting of two gamma heavy chains and two kappa or lambda light chains. The C region of the H chain of the molecule consists of three domains. Inter-chain disulfide linkages between the Cgamma1 and Cgamma2 domains stabilize the structure and define the hinge region of the molecule. IgG is the dominant immunoglobulin in extracellular fluids and is the only immunoglobulin transported across the placenta, and directly acts as an opsonin.

There are four subclasses of IgG, each of which displays unique antigenic determinants on the C region of the H chains. The approximate proportion of each subclass in blood is IgG1, 70%; IgG2, 20%; IgG3, 8%; and IgG4, 2%. The antibody specificities are distributed in somewhat specific patterns in each subclass. Neutralizing antibodies to protein toxins are mostly found in IgG1, antibodies to polysaccharides in IgG2, and antibodies to viruses in IgG3.

IgM: IgM is a pentamer of 4-chain units that are bound to a separate peptide called the J chain. IgM molecules consist of μ H chains and kappa or lambda L chains. Monomeric IgM is the principal antigen receptor on B cells. IgM is found principally in blood, but also occurs in external secretions. It binds most efficiently the C1q subunit of the first component of complement (vide infra), and is the first immunoglobulin expressed in B cell development.

IgA: IgA consists of a heavy chains and kappa or lambda light chains. There are two principal molecular forms of IgA, monomers whose basic structure and numbers of domains are similar to IgG and dimers that bind to J chains. Monomeric IgA, the second most common immunoglobulin in adult serum, is primarily produced by plasma cells in the bone marrow, whereas dimeric IgA, the dominant immunoglobulin in external secretions, is produced by plasma cells at mucosal sites.

Dimeric IgA is complexed and transported with a secretory component to form secretory IgA. Dimeric IgA binds to polymeric immunoglobulin receptors (secretory component) on the basolateral membranes of epithelial cells; the complex is internalized and transported across the cells in an endocytic vesicle to the apical pole of the cell where it is secreted as secretory IgA. The addition of a secretory component not only facilitates the transport of dimeric IgA, but protects the molecule from proteolysis.

There are two subclasses of IgA, IgA1 and IgA2. IgA1 predominates in the blood; there is an equal distribution of the two subclasses in external secretions. IgA2 is more resistant than IgA1 to bacterial IgA proteases that attack the hinge region of the molecule.

IgD: IgD is a monomeric four-chain polypeptide structure that is similar to IgG but its heavy chain (delta) is unique. Although this protein is expressed along with monomeric IgM on mature B cells, only small amounts of it are found in extracellular fluids.

IgE: IgE is also a four-chain polypeptide structure that is similar to IgG, but its heavy chain (epsilon) is distinct. Only trace amounts of this immunoglobulin are found in serum. IgE binds avidly to circulating blood basophils and mast cells in the submucosal sites and the skin. Cell-bound IgE antibodies defend against tissue parasites and initiate the pathogenesis of immediate hypersensitivity by triggering the release of low-molecular weight vasoactive compounds, including histamine, leukotrienes, and platelet-activating factor and certain proinflammatory cytokines such as TNF-alpha and IL-5, once they are cross-linked by antigens.

Sequence of Antibody Formation

Initial exposure to an antigen results in the production of low affinity antibodies, but continued exposure to antigen leads to the production of high affinity antibodies. In the primary antibody response (the first immunization), B cells are activated to produce IgM antibody. By 3-5 days, specific antibodies, mainly of the IgM isotype, appear in the serum and the concentration (titer) increases until a peak is reached in 10-14 days.

Antibody titers then fall to preimmunization levels after some weeks. Upon reimmunization, there is a more rapid and extensive development of antibody-producing cells in regional lymph nodes, and many of them undergo an isotype switch to produce IgG or other immunoglobulin classes of specific antibodies. As a result, in most cases following re-immunization, serum antibodies are primarily IgG and have a greater affinity for antigens; also, the antibody titers are higher and persist for much longer periods.

Antibody Binding to Antigen

There are a number of important consequences of antibody binding to antigens, depending upon the nature of the antigen. These include the neutralization of adherence sites or toxins from bacteria, the formation of opsonins, and the activation of the classical pathway of the complement system for that purpose or to create other bioactive factors that enhance inflammatory reactions.

(1) In the case of IgM, antigen-antibody complexes are created that most efficiently activate the classical pathway of the complement system and thereby lead to the formation of functional complement fragments including opsonins that facilitate the removal of the complexes by the RES.

(2) IgG antibodies, which are the dominant immunoglobulins in extracellular fluids, neutralize toxins and viruses, opsonize particles for ingestion by phagocytes, or when complexed to antigens, activate the classical pathway of complement.

(3) Secretory IgA antibodies defend mucosal sites by binding toxins and preventing adhesion of microbial pathogens.

(4) IgE antibodies on the surface of mast cells and basophils play an important role in defense against parasites and development of immediate hypersensitivity as previously noted.

Genetic Basis of Antibody Diversity

Specific antibodies are generated as a consequence of immunoglobulin gene rearrangement, i.e., recombination of V, D, J, and C genes. The immune system generates millions of different antibody molecules from the pool of V genes. Separate sets of V genes encode the variable domains of immunoglobulin H and L chains. The two chains are produced separately, but the mechanisms by which their diversity is achieved are similar in principle.

Light Chain Formation: Most antibody molecules use the kappa light chain. The kappa gene cluster consists of several hundred (~300) VL genes; a few J genes (~4) and one C gene. These germline genes are tandemly arranged on the chromosome and are transcriptionally inactive. As the B cell matures, genes are arranged (recombined) so that one V gene is joined to a J gene, and the rearranged VJ segment together with the C gene is transcribed.

The portion of DNA between the joined segments is deleted, and the transcripts are processed by splicing to produce the messenger RNA for the L chain. kappa chains are encoded by a separate cluster of V, J and C genes, but the rearrangement and transcription are similar to that of the lambda chain. Any given B cell uses only one type of L chain to produce the immunoglobulin molecule. The L chain combines with the H chain during their transport from polyribosomes to the membrane.

Heavy Chain Formation: The H chain gene system has a design that is similar to that of light chain, but is slightly more complex. In addition to ~ 1,000 VH genes, there are > 10 D genes and ~ 4 J genes. Furthermore, this genetic cluster has nine C genes that encode different immunoglobulin isotypes. The mature B cell rearranges its immunoglobulin genes, joins them together, and deletes the DNA between the joined segments. The rearranged VDJ gene segment is transcribed together with a Cμ or Cdelta gene, and this long transcript is spliced into VDJCμ or VDJCdelta messages resulting in the expression of IgM and IgD, respectively, on the B cell surface. Both immunoglobulin molecules use the same VDJ segment and, therefore, possess the same immunological specificity. The B cell is now ready to bind to a specific antigen and become further differentiated.

Generation of Antibody Diversity

Antibody diversity is generated by the following mechanisms.

Immunoglobulin Gene Rearrangements:

(1) The joining of various V, D and J genes is entirely random that results in ~ 50,000 different possible combinations for VDJ(H) and ~ 1,000 for VJ(L). Subsequent random pairing of H and L chains brings the total number of antibody specificities to $\sim 10^7$ possibilities.

(2) Diversity is further increased by the imprecise joining of different genetic segments.

(3) Rearrangements occur on both DNA strands, but only one strand is transcribed (allelic exclusion).

(4) Only one rearrangement occurs in the life of a B cell because of irreversible deletions in DNA. Consequently, each mature B cell maintains one immunologic specificity and is maintained in the progeny or clone.

This constitutes the molecular basis of the clonal selection; i.e., each antigenic determinant triggers the response of the pre-existing clone of B lymphocytes bearing the specific receptor molecule. It also follows that deletion of the B cell clone results in immunologic unresponsiveness to the antigen.

Somatic Mutations: This mechanism leads to a fine-tuning of the antibody specificity after immunization. Rearranged VDJH, and VJL genes in the B cells are uniquely susceptible to point mutagenesis by enzymes that become activated following stimulation of the cell by antigen. The clonal progeny of an antigen-driven B cell thus produce antibodies that may differ in one or more amino acid positions in the regions of the protein that are responsible for antigen binding. Cells producing the mutant antibody with highest affinity for the antigen are preferentially stimulated and thus eventually dominate the response. Therefore, antibodies produced after repeated immunization commonly display numerous point mutations (derived by somatic mutations in B cells found in peripheral lymphoid organs) and have higher affinities for antigens (affinity maturation), as compared to antibodies produced in the primary immune response.

Antibody Function: Antibody molecules perform a number of important functions that are necessary for mounting an effective immune response against microbial pathogens. CH region genes encode the biological functions of immunoglobulins. For example, IgM and IgG bind to the C1q subunit of C1, IgG crosses the placental barrier to the fetal circulation, and polymeric immunoglobulins, particularly dimeric IgA, are transported across epithelial cells into mucosal secretions.

To accomplish these functions, B cells switch their immunoglobulin isotype. The VDJ genes which are associated with Cμ or Cdelta, which are the original constant genes expressed in mature B cells, become associated with another C gene. This has been termed the isotype switch, because the C gene determines the antibody isotype. The switch is accomplished by genetic recombination, whereby the VDJ gene segment is transferred from the Cμ/Cdelta junction onto another C region gene downstream (Fig. 11). Because the Cμ/Cdelta and other interposed genes are deleted, the switch is irreversible. The new antibody maintains the same L chain and the same VH region (encoded by VDJ), but has new properties determined by the acquired C gene. The isotype switch mechanism is promoted by physical interactions between T and B cells (for example, the binding of CD40 on B cells to its ligand on T cells) and by specific cytokines from T cells (for example, IL-5 and IL-10 promote IgA production; IL-4 promotes IgE production).

Furthermore, each antibody molecule may exist in either a membrane-bound or secreted form. Every C gene contains a 3' sequence encoding the hydrophobic cytoplasmic tail of the H chain, so that the immunoglobulin molecule produced by the B cell is inserted in the surface membrane to function as the receptor for antigen. When the B cell differentiates into a plasma cell, an enzyme is activated that modifies the RNA transcript. Consequently, the translated protein ends with a hydrophilic peptide and is secreted from the cell.

TcR

The specific receptor for antigen on T lymphocytes, the TcR, is a heterodimeric protein with motifs that are similar to immunoglobulin molecules, but whose structure is encoded by a different set of V, J, D, and C genes. Moreover, T cells consist of two subsets carrying different receptors, that have been designated alpha/beta and gamma/delta.

A minority of T cells express a TcR consisting of gamma and delta chains and those cells are primarily $CD4^+$. These chains are encoded by very few genes; the gamma/delta repertoire is accordingly very limited. The gamma gene cluster consists of seven V genes, two J genes, and four C genes. The delta genes are interspersed within a gene locus that appears to include 10 V genes, two D genes, two J genes, and one C gene. Mature gamma/delta T cells seem to migrate primarily to mucosal and cutaneous tissues. The functions of gamma/delta T cells are not yet understood. Moreover, the recognition of antigen by gamma/delta T cells are not MHC-restricted.

TcR alpha/beta

Most T cells express the alpha/beta type of TcR. The genomic organization of the alpha and beta genes is more complicated than that of the immunoglobulin genes. Indeed, although a locus genes are interspersed with genes for the delta TcR (vide infra), alpha and beta genes are rearranged and expressed at different times and on different T lymphocytes.

The smaller alpha chain is encoded in a gene cluster consisting of ~100 V genes, ~50 J genes (a high number compared to immunoglobulin J genes) and one C gene. The alpha chains of various binding specificities are generated by a random genetic recombination of one V and one J gene, which are then joined with Calpha, by a mechanism analogous to that of the immunoglobulin L chain. The heavier beta chain is encoded by ~ 30 V genes, two D genes, >10 J genes and two C genes. The random joining of one of each V, D and J genes and their rearrangement to one C gene is similar to the process described for immunoglobulin genes. Rearranged VJCalpha and VDJCbeta DNA encodes the alpha and beta chain transcripts, respectively.

TcR genes are rearranged as lymphocytes mature in the thymus. Mature T cells, which are released from the thymus, are irreversibly committed to recognize one specific antigenic epitope in complex with self-MHC molecule.

Generation of TcR Diversity: The combinatorial diversity of TcR is greatly increased by junctional diversity, i.e., the variability of the junctions between different VDJ genes. New nucleotide base pairs are often added at the junction. Indeed, the junctional diversity of TcR is several orders of magnitude greater than that of an immunoglobulin gene. On the other hand, rearranged TcR genes are not subject to somatic mutations that contribute significantly to the generation of antibody diversity. The lack of somatic mutations appears to be related to the fact that the alpha/beta T cells always recognize a complex of antigenic fragment with the self MHC molecule. The receptor mutation could divert the specificity towards self molecules.

CD3 COMPLEX

The TcR cannot bind to soluble antigens but they recognize antigenic peptides bound to MHC molecules (i.e. class I or class II). Even after TcR binds MCH-peptide complex, it cannot transmit optimal signal necessary for T cell activation. Intracellular signalling of T cells requires non-covalent association of TcR with cell surface CD3 complex. The CD3 complex consists of four transmembrane peptides designated gamma, delta, epsilon, and zeta. The CD3 complex itself does not recognize the antigen and does not have variable domains. However, the CD3 complex transmits the biochemical signals generated by the TcR/antigen/MHC interaction on the surface that lead to lymphocyte activation.

MAJOR HISTOCOMPATIBILITY COMPLEX (MHC)

General Features: These molecules play a very important role in the recognition of self and non-self antigens by T cells. The MHC consists of a cluster of >100 genes on chromosome 6 that encode a number of biologically important molecules (Fig. 9). These molecules are responsible for the rejection of tissue grafts by genetically disparate individuals, as the name histocompatibility indicates. These molecules present antigens to T lymphocytes; govern interactions between T cells, B cells and accessory cells; and control the intrathymic development of the TcR repertoire against foreign antigens (positive selection) and against self (negative selection) Human MHC protein products are called human leukocyte antigens (HLA).

Genes and Structures: The two most important HLA glycoproteins are designated as class I and class II molecules.

MHC Class I/II Molecules: MHC class I molecules are ubiquitous on somatic cells whereas MHC class II molecules are restricted to monocytes, macrophages, dendritic cells, B cells, Langerhans cells, keratinocytes, activated T cells and certain types of epithelial cells. MHC class I molecules have three extracellular domains (alpha1, alpha2 and beta1), and a cytoplasmic tail. In contrast, MHC class II molecules have four extracellular domains (alpha1, alpha2, beta1 and beta2).

Three genes encode three independently expressed MHC class I molecules: HLA-A, -B and -C. Each gene contains three exons for the domains 1, 2 and 3. The MHC class II cluster, HLA-D, also contains three distinct genes, DP, DQ and DR, each of which has a separate set of exons for the alpha and beta chain.

MHC Alleles: An important aspect of the HLA gene system is its polymorphism. Each gene, MHC class I (A, B and C) and MHC class II (DP, DQ and DR) exists in different forms, or alleles. HLA alleles are designated by numbers and subscripts. For example, two unrelated individuals may carry class I HLA-B, genes B5, and Bw41, respectively. Allelic gene products differ in one or more amino acids in the alpha and/or beta domain(s). Large panels of specific antibodies are used to type HLA haplotypes of individuals using leukocytes that express class I and class II molecules. HLA typing is used for matching donors and recipients for organ/tissue transplantation and to predict the risk of certain diseases . In addition, the polymorphism of HLA genes has major implications for the function of class I and class II molecules (vide infra).

Role in Antigen Presentation: MHC molecules are required for antigen presentation to T cells. Peptides associated with MHC class I and class II molecules are recognized by $CD8^+$ and $CD4^+$ T cells, respectively. Foreign protein antigens are taken up by various types of cells in the body, internalized and subjected to enzymatic degradation called antigen processing. Antigenic peptide fragments bind to MHC class II molecules and are then transported to the cell surface. This MHC/antigen complex is recognized by the TcR on $CD4^+$ T cells. A $CD4^+$ T cell activated by an appropriate class II/peptide antigen complex on an antigen-presenting cell, such as a B cell or macrophage, may become a helper cell for antibody or cell-mediated immune responses.

A different scenario is found for viral antigens in infected cells or tumor antigens. These antigens are processed to fragments, which are expressed in association with the class I molecule. The MHC class I/antigen complex is recognized via the TcR by $CD8^+$ T lymphocytes, which become activated and differentiate into CTLs that destroy infected cells.

Variable domains of MHC class II molecules encoded by some allelic genes may be unable to bind a given antigenic peptide and thus fail to present the peptide to antigen-specific T cells. As a result, an immune response to this antigen cannot be mounted. Because of the association of high and low responses to a specific antigen with particular MHC class II alleles, MHC genes have been termed immune response genes. HLA typing reveals that individuals carrying certain alleles are at a higher risk of developing diseases such as ankylosing spondylitis, myasthenia gravis or type I diabetes mellitus. It is likely that this association reflects an underlying immunopathologic reaction involving MHC class I/II molecules or an association with other genes in the MHC.

MHC products control the selection of the immune repertoire of T lymphocytes. T cells interact with MHC class I/II molecules during their maturation in the thymus. This interaction kills immature cells whose TcR have a high affinity for self-MHC or for an MHC/self-protein complex by a mechanism called apoptosis or programmed cell death. Potentially autoreactive T cells would be eliminated in this fashion (negative selection). Furthermore, the selection process by MHC molecules determines the T cell repertoire of the individual against various foreign antigens (positive selection). This negative and positive sorting of T cells is called thymic selection.

T Cell Activation: Although, the presentation of antigen in the context of MHC molecules is essential for T cell recognition of peptide antigens, interactions between the MHC-bound peptide and TcR and the MHC class I or class II molecules, respectively, with CD8 or CD4 is not sufficient to activate T cells. Other ligands on antigen-presenting cells and their receptors on T cells are required to complete the process. These ligand-receptor interactions include the ligands ICAM-1, LFA-3, and B7-1/2 on antigen-presenting cells binding to their receptors LFA-1, CD2, and CD28/CTLA-4, respectively, on T cells. These and other counterstructures for B7-1 and B7-2 appear to precisely control the extent of T cell activation.

RECOGNITION OF SELF AND IMMUNE TOLERANCE

Self Tolerance

The immune system has evolved to distinguish between self and non-self antigens and to largely eliminate self-reactive lymphocytes. Because the repertoire of immune specificities is vast and largely random, it is not surprising that many nascent lymphocytes possess receptors for self-antigens. The mechanism of intrauterine tolerance is not well understood, but much has been learned about the mechanisms for excluding or inactivating self-reactive lymphocytes, particularly by using the model of experimentally induced immune tolerance to foreign antigens.

When an antigen is introduced into immunologically immature newborn animals, they may, upon reaching maturity, become unresponsive to immunization with that antigen (neonatal tolerance). This immunological tolerance is characterized by the absence of both antibody and cell-mediated responses, and it is specific for the original antigen.

Subsequent experiments revealed that the induction of antigen-specific tolerance is not always restricted to immature organisms. Unresponsiveness can also be induced in adults by using relatively higher doses of soluble antigen (high dose tolerance).

The induced state of unresponsiveness to the antigen is sometimes accompanied by the appearance of suppressor T cells that actively and specifically inhibit the responses of B and T cells. Recent studies also reveal that $IgM^{+}IgD^{-}$ B cells and mature T lymphocytes may be directly inactivated by small doses of antigen in vitro (low dose tolerance). In that model, short exposure of lymphocytes to the antigen, either at a critical concentration or in a certain modality, leads to an inactivation rather than a stimulation of the cells.

Collectively, the experiments on tolerance induction demonstrate that the unresponsiveness to self is likely to be achieved at several levels. During normal development, the self-reactive lymphocyte clones may be inactivated or deleted by exposure to self macromolecules during the early stages of maturation in the thymus.

The autoselection is dependent upon MHC class I molecules for $CD8^{+}$ T cells and class II molecules for $CD4^{+}$ T cells. Those cells that are not eliminated and reach their full immunological potential may be inactivated, when self molecules are presented to these cells at high concentrations or in a form that is tolerogenic rather than immunogenic. Also, it is possible that some self-reactive lymphocytes are suppressed by other regulatory cells, such as $CD8^+$ suppressor T cells.

Autoimmunity

The failure of any of the mechanisms involved in self recognition and elimination or down regulation of self-reactive clones may result in autoimmunity. Autoimmune disorders in genetically prone individuals may be generated by (a) changes in the expression of self macromolecules or alterations in their presentation to lymphocytes, (b) release of sequestered self-antigens into the circulation, or access of immunogens to normally immunologically privileged sites, and (c) alterations in lymphocyte maturation and immune regulation. In addition, foreign antigens such as bacteria and viruses that cross-react with self antigens may augment or initiate any of the above mechanisms.

The Complement System

The complement system consists of a group of glycoproteins in the extracellular space that can be stimulated in a cascading fashion to produce biologically active fragments that either directly attack foreign substances or enhance the functions of certain types of inflammatory leukocytes. The complement system consists of two recognition-stimulation pathways that are designated as the classical and alternative pathways, either of which may lead to the formation of a cell membrane attack complex.

The Classical Pathway

The classical pathway of the complement system may be activated by antigen-antibody complexes of the IgG, IgG3, or IgM isotypes by their binding to the C1q subunit of the first component of complement. Consequently, the C1qrs subunits of C1 form an esterase that cleaves the next component, C4, to two fragments, the larger of which, C4b, binds covalently to hydroxyl or amino groups on cellular membranes. The next component, C2, after binding to C4b is partially digested by C1s esterase to form C2b. The resultant membrane-bound complex, C4b2a, is an enzyme (C3 convertase) that cleaves C3 into two biologically active fragments, C3a and C3b.

The Alternative Pathway

The alternative pathway of the complement system is activated independently of antigen-antibody complexes. The major exogenous activators of the pathway are microbial agents and their products. The major components of the pathway are the serum protein factors B, D, and P (properdin). A small amount of C3 in the fluid phase, which normally is spontaneously activated, interacts with factor B to form C3Bb, which cleaves other C3 molecules to form C3b. C3b in turn attaches to surfaces and binds factor B. The resultant C3bB is then cleaved by factor D to form C3bBb, the C3 convertase of the alternative pathway. That enzyme is distinct from the one generated from the classical pathway but serves the same purpose. This complex then is stabilized by factor P.

The binding of C3 to factor B is prevented, particularly in the fluid phase, by a regulatory molecule, factor H. The more vigorous activation of this pathway occurs when the host is exposed to microorganisms that are poor in sialic acid. In those circumstances, the binding of factor B to C3 is favored, and the activation of the alternative pathway is not readily inhibited by factor H. Therefore, more C3b is generated and a positive amplification loop that generates more C3bBb (C3 convertase) is created.

In contrast, sialic acid-rich encapsulated microorganisms such as Streptococcus pneumoniae, Haemophilus influenzae, and Niesseria meningitides are incapable of activating the alternative pathway and require binding to specific IgG or IgM antibodies to activate the classical pathway and generate the C3b for phagocytosis and the formation of the membrane attack complex.

The receptors for activated complement fragments are (1) CR1, principally on phagocytic cells for C3b; (2) CR2, principally on B cells for a fragment called C3d (receptor for EBV); and CR3 (Mac-1), on phagocytic and NK cells for inactivated C3b (C3bi) and C3d-g fragments.

The Membrane Attack Complex

The activation of the complement system eventually leads to the formation of the membrane attack complex that consequently lyses cells. The membrane attack complex is formed in the following manner. As a result of the formation of C3b, C5 is cleaved into two fragment, C5b and C5a. The larger fragment, C5b, combines with C6 and the complex attaches to the cell surface, where it forms the foundation for the sequential binding of C7, 8 and 9, e.g., the membrane attack complex. C3b and its degradation product, C3bi, are opsonins. C3a and C5a are chemotaxins and anaphylotoxins; C5a is the more potent of the two factors.

Once the membrane attack complex is formed, discrete holes are created in the surface membranes of the target cells. Consequently, extracellular fluid accumulates in the target cell, eventually leading to its lysis.

DEFENSE AGAINST INFECTIONS

Cutaneous and Mucosal Defense

The first line of defense against most potential pathogens is the skin and mucous membranes. In addition to anatomic barriers, certain protective biochemical agents are produced at mucosal sites. These include simple chemicals such as acids and bases, and macromolecular proteins, including lysozyme, lactoferrin, secretory IgA antibodies, and interferons.

The genesis of secretory IgA antibodies is as follows. Under the influence of IL-5, IL-6, and IL-10, B cells bearing surface IgM in Peyer's patches and in the submucosa of the tracheobronchial tree switch to IgA-bearing cells. When the surface antibodies of these altered cells combine with a specific antigen, the cells are stimulated to migrate through afferent lymphatics to the regional lymph nodes and then through efferent lymphatic channels into the vascular circulation.

They then home to submucosal sites in the upper small intestine, or to the respiratory system, where they differentiate into plasma cells that secrete large amounts of specific dimeric IgA antibodies and are transported across epithelial cells to the lumen by secretory component, as previously described. The resultant secretory IgA is particularly well suited to mucosal sites since it is more resistant than other types of immunoglobulins to the digestive processes of the alimentary tract. These antibodies protect by complexing adherence structures and toxins from microbial pathogens.

Activation of Local Immunity

The second line of defense consists of local factors and leukocytes that are activated or recruited to the site of microbial invasion. These local elements of defense include the coagulation system, the fibrinolytic system, kallikrein, the complement system, resident macrophages, and elicited inflammatory cells.

Activation of Systemic Immunity

If the pathogen is able to overcome the first two lines of defense, systemic acquired responses are marshalled to prevent further invasion and damage. This third line of defense includes intracellular killing by circulating phagocytes, stimulation of monokine production, interleukin production by T cells, production of circulating antibodies by plasma cells in regional lymph nodes and the spleen, intravascular activation of the complement system, and phagocytosis of opsonized pathogens by cells of the RES. Cytotoxic mechanisms directed against ingested microbes or infected cells play a major role in defense.

Unless the microbial inoculum is overwhelming, unusually virulent, or the host defenses are compromised, the infection should be contained and finally obliterated via a combination of local and systemic responses. At the same time local fibroblasts and epithelial cells proliferate, the tissue becomes more vascularized, and debris is removed by local tissue phagocytes. The inflammatory reaction abates and the tissue heals.

Diseases Due to Immune Responses to Infectious Agents

Five major types of immune responses to infecting agents may lead to disease.

(1) Circulating immune complexes formed from microbial antigens such as hepatitis B virus bound to IgM or IgG

antibodies, may deposit in skin, synovia, or glomeruli and elicit inflammation by activating the classical pathway of complement.

(2) Invading microorganisms may give rise to antibodies that cross-react with autoantigens. For example, antibodies produced against Group A, beta-hemolytic streptococci in patients with rheumatic fever often react against sarcolemmal antigens in cardiac muscle.

(3) Vasoactive compounds may be released into local tissues or the systemic circulation because of activation of the alternative pathway of complement by certain bacteria deficient in sialic acid such as Salmonella.

(4) Cytokines such as TNF-alpha, IL-1, and IFN-gamma released during infection from stimulated macrophages and T lymphocytes, may lead to fever, dysregulate nutritional pathways, and contribute to the vascular instability seen in sepsis.

(5) Finally, delayed hypersensitivity reactions that damage surrounding tissues occur in indolent infections such as tuberculosis by the formation of granulomas consisting of activated macrophages and cytotoxic T cells.

Ontogeny of Immunity

There is an orderly development of the immune system during the intrauterine period. Pluripotent stem cells appear first in the yolk sac, then in the fetal liver, and finally the bone marrow. Neutrophils, monocytes, and macrophages are produced during fetal life, but the mononuclear phagocytes do not mature until after birth.

An epithelial thymus appears during the first few fetal weeks and then becomes populated with lymphocytes. Mature T and B cells appear in the blood soon thereafter, but they are largely not activated. Furthermore, IgG antibodies are usually not produced until after birth, and IgG antibodies to polysaccharide antigens do not appear until ~ 2 years of age. In addition, there are developmental delays in the production of certain cytokines, including GM-CSF, IL-10, TNF-alpha and IFN-gamma.

Neonates have as many B and T cells in the peripheral blood as do adults, these cells in the peripheral lymphoid organs are not as well developed because of the paucity of prenatal antigenic stimuli. As antigen stimulation occurs, the T and B cell zones of the peripheral lymphoid organs are progressively populated and the products of these stimulated cells, such as antibodies, begin to appear. The sequence of immunoglobulin production is as follows: IgM production occurs first and is then followed by IgG and IgA. Systemic IgG antibodies to polysaccharides are not produced, however, until the child is 2 to 2.5 years old. The secretory component is produced at birth, but the main immunoglobulin in external secretions in the first few weeks of postnatal life is IgM. Subsequently, secretory IgA becomes the dominant immunoglobulin at mucosal sites.

MATERNAL IMMUNOLOGIC AGENTS TRANSFERRED TO THE RECIPIENT INFANT

The mother transmits immune factors to the offspring both through the placenta and milk. Large quantities of IgG are transmitted via the placenta, whereas other immunoglobulin isotypes are not. Consequently, virtually all IgG in neonatal blood is of maternal origin, the concentration of IgG in umbilical cord blood is somewhat higher than in adults, and the levels of other immunoglobulin isotypes are exceptionally low. Low concentrations of some factors such as IgG and secretory IgA antibodies are also transmitted via amniotic fluid, but little is known about their in vivo effects upon the fetal mucosal immune system.

An array of host resistance factors are transmitted to the infant in human milk, including leukocytes, secretory IgA, lactoferrin, lysozyme, and oligosaccharides and glycoconjugates that are receptor analogs for microbial adhesins and toxins. In addition to those antimicrobial factors, anti-inflammatory agents, and immunomodulating agents including TNF-alpha, TGF-beta, IL-1beta, IL-6, IL-8, IL-10, G-CSF, and M-CSF. These factors are designed to act at mucosal sites and to protect by noninflammatory mechanisms. Since the endogenous production of these agents is incompletely developed in early infancy and they are scarce in cow's milk or other substitute feedings, it is not surprising that breastfeeding increases resistance to gastrointestinal and respiratory infections, allergic diseases, and certain inflammatory diseases that occur much later in childhood.

Immune Deficiencies

Immune deficiencies may be due to genetic or acquired defects, and these defects lead to increased risks to certain infectious diseases depending upon the specific immune defects. Much of the basic information concerning the development and function of the immune system has been learned from investigations of inherited, congenital, and acquired defects of the system. Examples of the principal defects that have lead to an elucidation of the immune system are as follows.

Genetic Defects

X-Linked Agammaglobulinemia: In this immunoglobulin deficiency disease, there is a genetic defect in the development of B cells from pre-B cells in the bone marrow. The defect is due to mutations in the gene that encodes for B cell tyrosine kinase. Consequently, the B cells are not produced. Because of the block in the development of B cells, germinal centers, plasma cells, and specific antibodies are profoundly reduced. The rest of the immune system is normal. Affected individuals are unusually susceptible to infection by virulent encapsulated respiratory bacteria and enteroviruses. These patients benefit greatly from intravenous infusions of human IgG.

Hyper-IGM Antibody Deficiency: A second X-linked defect in antibody formation, the hyper-IgM antibody deficiency, is characterized by a block in immunoglobulin class switching. Consequently, IgM (and IgD) antibodies are produced, but IgA and IgG antibodies are not.

These patients are also unusually susceptible to infection by virulent encapsulated respiratory bacteria. The disease is due to mutation in the gene that encodes for CD40 ligand (CD39) on T cells. Because of the defect, T and B cell interactions are insufficient for immunoglobulin class switching. These patients also benefit greatly from intravenous infusions of human IgG.

Severe Combined Immunodeficiency (SCID): The most common type of SCID is due to stop codon defects in the X-chromosome gene that encodes for the gamma-chain that is common to IL-2, IL-4, IL-7, IL-9, and IL-15 receptors. A more moderate combined immunodeficiency disease has been reported and is due to a missense point mutation in a part of the gene that encodes for the cytoplasmic region of that gene. In addition, other defects reported to cause SCID involve an autosomal recessive defect in the formation of adenosine deaminase, a defect in the formation of CD3, a defect in the post TcR-CD3 receptors' signalling, and deficiency in the formation of IL-2.

Patients with these diseases display few T lymphocytes, decreased T cell functions, poor antibody formation, and variable numbers of B cells and serum concentrations of immunoglobulins. As a consequence of the deficiencies in T cells, these patients are very susceptible to opportunistic pathogens, including Candida albicans, Salmonellae, Pneumocystis carinii, Cytomegalovirus, and Varicella zoster virus. Patients with SCID usually die before the age of two years, unless definitive immunologic interventions are instituted.

Specific treatments have been devised for patients with SCID. Many patients have been treated successfully with bone marrow transplants to supply normal stem cells. Patients with adenosine deaminase deficiency may also be managed by infusing the enzyme packaged with polyethylene glycol. Recently, some patients with adenosine deaminase deficiency have been successfully treated with gene therapy. Although the beneficial effects have not been permanent, nevertheless, they are encouraging.

Two major intrinsic defects in the function of phagocytic cells have been recognized. The first one is an autosomal recessive defect in the formation of the common beta-subunit of the family of adherence glycoproteins (integrins). The deficiency interferes with the ability of these leukocytes to adhere to the surface of endothelial cells. Consequently, the motility of these cells on two-dimensional surfaces is impaired. Thus, this defect results in bacterial infections in interstitial sites, such as in the skin and periodontium.

The second disorder, chronic granulomatous disease, was the first recognized genetic defect of the function of phagocytic cells. The disease is X-linked in ~ 70% of affected patients. In those cases, the gene for the gp90 protein subunit of cytochrome b_{558} is abnormal. Consequently, the protein is not produced, the cytochrome does not persist, and intracellular killing is impaired. Less frequently (~ 3% of cases), the disease is due to a defect in the autosomal gene for the lower molecular weight subunit of the heterodimer, p22phox. Autosomal defects in genes for cytoplasmic proteins that stabilize the cytochrome have also been recognized.

In these disorders, phagocytes are unable to mount a respiratory burst and therefore are unable to produce toxic oxygen compounds, such as hydrogen peroxide, which are required for intracellular killing of catalase-positive microorganisms such as Candida albicans, Escherichia coli, and Serratia species. The failure to kill catalase-positive microorganisms occurs because the microbial agents do not supply the oxygen substrates required for intracellular killing. In contrast, these dysfunctional neutrophils kill catalase-negative microorganisms such as the streptococci since those microorganisms bring hydrogen peroxide into the phagolysosome.

Acquired Defects

Protein-Calorie Malnutrition: Protein-calorie malnutrition is the leading cause of immune deficiency in the world. Protein-calorie malnutrition leads principally to a profound deficiency in the production and function of T cells, rendering the victim susceptible to many of the opportunistic infections that occur in genetic T cell deficiencies. With increasing protein-energy deficiency other parts of the immune system are also affected. Specific nutrient deficits such as iron or vitamin A deficiency also depress certain parts of the immune system.

Certain types of infections temporarily depress parts of the immune system. For example, many acute viral infections suppress cellular immunity for several days to a few weeks, and serious bacterial infections inhibit the ability of neutrophils to respond to chemotactic agents. Furthermore in schistosomiasis, Th1 responses are accentuated and Th2 responses are suppressed. This leads to a decreased ability to form antibodies after antigenic challenges.

Malnutrition and infection interact to inhibit the immune system. As a result of malnutrition, the immune system becomes compromised. That leads to respiratory and gastrointestinal infections. Those infections may in turn further interfere with the immune system. Moreover, some infections further compromise nutritional status by increasing nutrient losses (intestinal malabsorption) and utilization (fever, caloric expenditure during sweating) or by interfering with normal nutrient metabolic pathways (cachectic effect of TNF-alpha). Consequently, immune function is further impaired.

Human Immunodeficiency Virus Infections: A second acquired immunodeficiency is due to human immunodeficiency virus (HIV) infection. This retrovirus infection was first encountered in homosexual males and individuals who were injecting illicit drugs or who received blood products contaminated with the virus. The infection has since spread to heterosexual populations by sexual transmission. The infection has reached epidemic proportions in developed as well as developing countries and continues to increase. The resultant acquired immune deficiency syndrome (AIDS) occurs because the virus infects and destroys $CD4^+$ T cells. The virus binds to the CD4 surface antigen on T cells and to the same or similar moiety on macrophages. Since $CD4^+$ T cells are essential for the genesis of cellular immunity and for orchestrating the function of many other parts of the immune system, a deficiency in these T cells increases the patient's susceptibility to opportunistic infections. The vast majority of such infected patients die after several years. No preventative immunizations or curative treatments are available for the infection at this time.

BACTERIOLOGY

Bacteria are single-celled microorganisms that lack a nuclear membrane, are metabolically active and divide by binary fission. Medically they are a major cause of disease. Superficially, bacteria appear to be relatively simple forms of life; in fact, they are sophisticated and highly adaptable. Many bacteria multiply at rapid rates, and different species can utilize an enormous variety of hydrocarbon substrates, including phenol, rubber, and petroleum. These organisms exist widely in both parasitic and free-living forms. Because they are ubiquitous and have a remarkable capacity to adapt to changing environments by selection of spontaneous mutants, the importance of bacteria in every field of medicine cannot be overstated.

The discipline of bacteriology evolved from the need of physicians to test and apply the germ theory of disease and from economic concerns relating to the spoilage of foods and wine. The initial advances in pathogenic bacteriology were derived from the identification and characterization of bacteria associated with specific diseases. During this period, great emphasis was placed on applying Koch's postulates to test proposed cause-and-effect relationships between bacteria and specific diseases. Today, most bacterial diseases of humans and their etiologic agents have been identified, although important variants continue to evolve and sometimes emerge, e.g., Legionnaire's Disease, tuberculosis and toxic shock syndrome.

Major advances in bacteriology over the last century resulted in the development of many effective vaccines (e.g., pneumococcal polysaccharide vaccine, diphtheria toxoid, and tetanus toxoid) as well as of other vaccines (e.g., cholera, typhoid, and plague vaccines) that are less effective or have side effects. Another major advance was the discovery of antibiotics. These antimicrobial substances have not eradicated bacterial diseases, but they are powerful therapeutic tools. Their efficacy is reduced by the emergence of antibiotic resistant bacteria (now an important medical management problem) In reality, improvements in sanitation and water purification have a greater effect on the incidence of bacterial infections in a community than does the availability of antibiotics or bacterial vaccines. Nevertheless, many and serious bacterial diseases remain.

Most diseases now known to have a bacteriologic etiology have been recognized for hundreds of years. Some were described as contagious in the writings of the ancient Chinese, centuries prior to the first descriptions of bacteria by Anton van Leeuwenhoek in 1677. There remain a few diseases (such as chronic ulcerative colitis) that are thought by some investigators to be caused by bacteria but for which no pathogen has been identified. Occasionally, a previously unrecognized diseases is associated with a new group of bacteria. An example is Legionnaire's disease, an acute respiratory infection caused by the previously unrecognized genus, Legionella. Also, a newly recognized pathogen, *Helicobacter,* plays an important role in peptic disease. Another important example, in understanding the etiologies of venereal diseases, was the association of at least 50 percent of the cases of urethritis in male patients with *Ureaplasma urealyticum* or *Chlamydia trachomatis.*

Recombinant bacteria produced by genetic engineering are enormously useful in bacteriologic research and are being employed to manufacture scarce biomolecules (e.g. interferons) needed for research and patient care. The antibiotic resistance genes, while a problem to the physician, paradoxically are indispensable markers in performing genetic engineering. Genetic probes and the polymerase chain reaction (PCR) are useful in the rapid identification of microbial pathogens in patient specimens. Genetic manipulation of pathogenic bacteria continues to be indispensable in defining virulence mechanisms. As more protective protein antigens are identified, cloned, and sequenced, recombinant bacterial vaccines will be constructed that should be much better than the ones presently available. In this regard, a recombinant-based and safer pertussis vaccine is already available in some European countries. Also, direct DNA vaccines hold considerable promise.

In developed countries, 90 percent of documented infections in hospitalized patients are caused by bacteria. These cases probably reflect only a small percentage of the actual number of bacterial infections occurring in the general population, and usually represent the most severe cases. In developing countries, a variety of bacterial infections often exert a devastating effect on the health of the inhabitants. Malnutrition, parasitic infections, and poor sanitation are a few of the factors contributing to the increased susceptibility of these individuals to bacterial pathogens. The World Health Organization has estimated that each year, 3 million people die of tuberculosis, 0.5 million die of pertussis, and 25,000 die of typhoid. Diarrheal diseases, many of which are bacterial, are the second leading cause of death in the world (after cardiovascular diseases), killing 5 million people annually.

Many bacterial diseases can be viewed as a failure of the bacterium to adapt, since a well-adapted parasite ideally thrives in its host without causing significant damage. Relatively nonvirulent (i.e., well-adapted) microorganisms can cause disease under special conditions - for example, if they are present in unusually large numbers, if the host's defenses are impaired, (e.g., AIDS and chemotherapy) or if anaerobic conditions exist. Pathogenic bacteria constitute only a small proportion of bacterial species; many nonpathogenic bacteria are beneficial to humans (i.e. intestinal flora produce vitamin K) and participate in essential processes such as nitrogen fixation, waste breakdown, food production, drug preparation, and environmental bioremediation. This textbook emphasizes bacteria that have direct medical relevance.

In recent years, medical scientists have concentrated on the study of pathogenic mechanisms and host defenses. Understanding host-parasite relationships involving specific pathogens requires familiarity with the fundamental characteristics of the bacterium, the host, and their interactions. Therefore, this section first presents with the basic concepts of the immune response, bacterial structure, taxonomy, metabolism, and genetics. Subsequent chapters emphasize normal relationships among bacteria on external surfaces; mechanisms by which microorganisms damage the host; host defense mechanisms; source and distribution of pathogens (epidemiology); principles of diagnosis; and mechanisms of action of antimicrobial drugs. These chapters provide the basis for the next chapters devoted to specific bacterial pathogens and the diseases they cause. The bacteria in these chapters are grouped on the basis of physical, chemical, and biologic characteristics. These similarities do not necessarily indicate that their diseases are similar; widely divergent diseases may be caused by bacteria in the same group.

Structure

All bacteria, both pathogenic and saprophytic, are unicellular organisms that reproduce by binary fission. Most bacteria are capable of independent metabolic existence and growth, but species of Chlamydia and Rickettsia are obligately intracellular organisms.

Bacterial cells are extremely small and are most conveniently measured in microns (10-6 m). They range in size from large cells such as Bacillus anthracis (1.0 to 1.3 μm X 3 to 10 μm) to very small cells such as Pasteurella tularensis (0.2 X 0.2 to 0.7 μm) Mycoplasmas (atypical pneumonia group) are even smaller, measuring 0.1 to 0.2 μm in diameter. Bacteria therefore have a surface-to-volume ratio that is very high: about 100,000.

Bacteria have characteristic shapes. The common microscopic morphologies are cocci (round or ellipsoidal cells, such as Staphylococcus aureus or Streptococcus respectively); rods, such as Bacillus and Clostridium species; long, filamentous branched cells, such as Actinomyces species; and comma-shaped and spiral cells, such as Vibrio cholerae and Treponema pallidum, respectively. The arrangement of cells is also typical of various species or groups of bacteria. Some rods or cocci characteristically grow in chains; some, such as Staphylococcus aureus, form grapelike clusters of spherical cells; some round cocci form cubic packets. Bacterial cells of other species grow separately. The microscopic appearance is therefore valuable in classification and diagnosis.

The higher resolving power of the electron microscope not only magnifies the typical shape of a bacterial cell but also clearly resolves its prokaryotic organization.

The Nucleoid

Prokaryotic and eukaryotic cells were initially distinguished on the basis of structure: the prokaryotic nucleoidthe equivalent of the eukaryotic nucleusis structurally simpler than the true eukaryotic nucleus, which has a complex mitotic apparatus and surrounding nuclear membrane. As the electron micrograph in figure above shows, the bacterial nucleoid, which contains the DNA fibrils, lacks a limiting membrane. Under the light microscope, the nucleoid of the bacterial cell can be visualized with the aid of Feulgen staining, which stains DNA. Gentle lysis can be used to isolate the nucleoid of most bacterial cells. The DNA is then seen to be a single, continuous, "giant" circular molecule with a molecular weight of approximately 3 X 109. The unfolded nuclear DNA would be about 1 mm long (compared with an average length of 1 to 2 μm for bacterial cells). The bacterial nucleoid, then, is a structure containing a single chromosome. The number of copies of this chromosome in a cell depends on the stage of the cell cycle (chromosome replication, cell enlargement, chromosome segregation, etc). Although the mechanism of segregation of the two sister chromosomes following replication is not fully understood, all of the models proposed require that the chromosome be permanently attached to the cell membrane throughout the various stages of the cell cycle.

Bacterial chromatin does not contain basic histone proteins, but low-molecular-weight polyamines and magnesium ions may fulfill a function similar to that of eukaryotic histones. Despite the differences between prokaryotic and eukaryotic DNA, prokaryotic DNA from cells infected with bacteriophage g, when visualized by electron microscopy, has a beaded, condensed appearance not unlike that of eukaryotic chromatin.

Surface Appendages

Two types of surface appendage can be recognized on certain bacterial species: the flagella, which are organs of locomotion, and pili (Latin hairs), which are also known as fimbriae (Latin fringes). Flagella occur on both Gram-positive and Gram-negative bacteria, and their presence can be useful in identification. For example, they are found on many species of bacilli but rarely on cocci. In contrast, pili occur almost exclusively on Gram-negative bacteria and are found on only a few Gram-positive organisms (e.g., Corynebacterium renale).

Some bacteria have both flagella and pili. The electron micrograph in figure below shows the characteristic wavy appearance of flagella and two types of pili on the surface of Escherichia coli.

Flagella

Structurally, bacterial flagella are long (3 to 12 μm), filamentous surface appendages about 12 to 30 nm in diameter. The protein subunits of a flagellum are assembled to form a cylindrical structure with a hollow core. A flagellum consists of three parts: (1) the long filament, which lies external to the cell surface; (2) the hook structure at the end of the filament; and (3) the basal body, to which the hook is anchored and which imparts motion to the flagellum. The basal body traverses the outer wall and membrane structures. It consists of a rod and one or two pairs of discs. The thrust that propels the bacterial cell is provided by counterclockwise rotation of the basal body, which causes the helically twisted filament to whirl. The movement of the basal body is driven by a proton motive force rather than by ATP directly. The ability of bacteria to swim by means of the propeller-like action of the flagella provides them with the mechanical means to perform chemotaxis (movement in response to attractant and repellent substances in the environment). Response to chemical stimuli involves a sophisticated sensory system of receptors that are located in the cell surface and/or periplasm and that transmit information to methyl-accepting chemotaxis proteins that control the flagellar motor. Genetic studies have revealed the existence of mutants with altered biochemical pathways for flagellar motility and chemotaxis.

Chemically, flagella are constructed of a class of proteins called flagellins. The hook and basal-body structures consist of numerous proteins. Mutations affecting any of these gene products may result in loss or impairment of motility. Flagellins are immunogenic and constitute a group of protein antigens called the H antigens, which are characteristic of a given species, strain, or variant of an organism. The species specificity of the flagellins reflects differences in the primary structures of the proteins. Antigenic changes of the flagella known as the phase variation of H1 and H2 occurs in Salmonella typhimurium.

The number and distribution of flagella on the bacterial surface are characteristic for a given species and hence are useful in identifying and classifying bacteria. For example, V cholerae has a single flagellum at one pole of the cell (i.e., it is monotrichous), whereas Proteus vulgaris and E coli have many flagella distributed over the entire cell surface (i.e., they are peritrichous). The flagella of a peritrichous bacterium must aggregate as a posterior bundle to propel the cell in a forward direction.

Flagella can be sheared from the cell surface without affecting the viability of the cell. The cell then becomes temporarily nonmotile. In time it synthesizes new flagella and regains motility. The protein synthesis inhibitor chloramphenicol, however, blocks regeneration of flagella.

Pili

The terms pili and fimbriae are usually used interchangeably to describe the thin, hairlike appendages on the surface of many Gram-negative bacteria and proteins of pili are referred to as pilins. Pili are more rigid in appearance than flagella. In some organisms, such as Shigella species and E coli, pili are distributed profusely over the cell surface, with as many as 200 per cell. As is easily recognized in strains of E coli, pili can come in two types: short, abundant common pili, and a small number (one to six) of very long pili known as sex pili. Sex pili can be distinguished by their ability to bind male-specific bacteriophages (the sex pilus acts as a specific receptor for these bacteriophages). The sex pili attach male to female bacteria during conjugation.

Pili in many enteric bacteria confer adhesive properties on the bacterial cells, enabling them to adhere to various epithelial surfaces, to red blood cells (causing hemagglutination), and to surfaces of yeast and fungal cells. These adhesive properties of piliated cells play an important role in bacterial colonization of epithelial surfaces and are therefore referred to as colonization factors. The common pili found on E coli exhibit a sugar specificity analogous to that of phytohemagglutinins and lectins, in that adhesion and hemagglutinating capacities of the organism are inhibited specifically by mannose. Organisms possessing this type of hemagglutination are called mannose-sensitive organisms. Other piliated organisms, such as gonococci, are adhesive and hemagglutinating, but are insensitive to the inhibitory effects of mannose. Extensive antigenic variations in pilins of gonococci are well known.

Surface Layers

The surface layers of the bacterial cell have been identified by various techniques: light microscopy and staining; electron microscopy of thin-sectioned, freeze-fractured, and negatively stained cells; and isolation and biochemical characterization of individual morphologic components of the cell. The principal surface layers are capsules and loose slime, the cell wall of Gram-positive bacteria and the complex cell envelope of Gram-negative bacteria, plasma (cytoplasmic) membranes, and mesosomal membrane vesicles, which arise from invaginations of the plasma membrane. In bacteria, the cell wall forms a rigid structure of uniform thickness around the cell and is responsible for the characteristic shape of the cell (rod, coccus, or spiral). Inside the cell wall (or rigid peptidoglycan layer) is the plasma (cytoplasmic) membrane; this is usually closely apposed to the wall layer. The latter shows the typical fracture planes seen in most Gram-negative bacteria, which are weak cleavage planes through the outer membrane of the envelope and extensive fracture planes through the bilayer region of the underlying plasma membrane.

Capsules and Loose Slime

Some bacteria form capsules, which constitute the outermost layer of the bacterial cell and surround it with a relatively thick layer of viscous gel. Capsules may be up to 10 μm thick. Some organisms lack a well-defined capsule but have loose, amorphous slime layers external to the cell wall or cell envelope. The a hemolytic Streptococcus mutans, the primary organism found in dental plaque is able to synthesis a large extracellular mucoid glucans from sucrose. Not all bacterial species produce capsules; however, the capsules of encapsulated pathogens are often important determinants of virulence. Encapsulated species are found among both Gram-positive and Gram-negative bacteria. In both groups, most capsules are composed of highmolecular-weight viscous polysaccharides that are retained as a thick gel outside the cell wall or envelope. The capsule of Bacillus anthracis (the causal agent of anthrax) is unusual in that it is composed of a g-glutamyl polypeptide. A plasma membrane stage is involved in the biosynthesis and assembly of the capsular substances, which are extruded or secreted through the outer wall or envelope structures. Mutational loss of enzymes involved in the biosynthesis of the capsular polysaccharides can result in the smooth-to-rough variation seen in the pneumococci.

The capsule is not essential for viability. Viability is not affected when capsular polysaccharides are removed enzymatically from the cell surface. The exact functions of capsules are not fully understood, but they do confer resistance to phagocytosis and hence provide the bacterial cell with protection against host defenses to invasion.

CELL WALL AND GRAM-NEGATIVE CELL ENVELOPE

The Gram stain broadly differentiates bacteria into Gram-positive and Gram-negative groups; a few organisms are consistently Gram-variable. Gram-positive and Gram-negative organisms differ drastically in the organization of the structures outside the plasma membrane but below the capsule: in Gram-negative organisms these structures constitute the cell envelope, whereas in Gram-positive organisms they are called a cell wall.

Most Gram-positive bacteria have a relatively thick (about 20 to 80 nm), continuous cell wall (often called the sacculus), which is composed largely of peptidoglycan (also known as mucopeptide or murein). In thick cell walls, other cell wall polymers (such as the teichoic acids, polysaccharides, and peptidoglycolipids) are covalently attached to the peptidoglycan. In contrast, the peptidoglycan layer in Gram-negative bacteria is thin (about 5 to 10 nm thick); in E coli, the peptidoglycan is probably only a monolayer thick. Outside the peptidoglycan layer in the Gram-negative envelope is an outer membrane structure (about 7.5 to 10 nm thick). In most Gram-negative bacteria, this membrane structure is anchored noncovalently to lipoprotein molecules (Braun's lipoprotein), which, in turn, are covalently linked to the peptidoglycan. The lipopolysaccharides of the Gram-negative cell envelope form part of the outer leaflet of the outer membrane structure.

The organization and overall dimensions of the outer membrane of the Gram-negative cell envelope are similar to those of the plasma membrane (about 7.5 nm thick). Moreover, in Gram-negative bacteria such as E coli, the outer and inner membranes adhere to each other at several hundred sites (Bayer patches); these sites can break up the continuity of the peptidoglycan layer.

The basic differences in surface structures of Gram-positive and Gram-negative bacteria explain the results of Gram staining. Both Gram-positive and Gram-negative bacteria take up the same amounts of crystal violet (CV) and iodine (I). The CV-I complex, however, is trapped inside the Gram-positive cell by the dehydration and reduced porosity of the thick cell wall as a result of the differential washing step with 95 percent ethanol or other solvent mixture. In contrast, the thin peptidoglycan layer and probable discontinuities at the membrane adhesion sites do not impede solvent extraction of the CV-I complex from the Gram-negative cell. The above mechanism of the Gram stain based on the structural differences between the two groups has been confirmed by sophisticated methods of electron microscopy. Moreover, mechanical disruption of the cell wall of Gram-positive organisms or its enzymatic removal with lysozyme results in complete extraction of the CV-I complex and conversion to a Gram-negative reaction. Therefore, autolytic wall-degrading enzymes that cause cell wall breakage may account for Gram-negative or variable reactions in cultures of Gram-positive organisms (such as

Staphylococcus aureus, Clostridium perfringens, Corynebacterium diphtheriae, and some Bacillus spp).

Peptidoglycan

Unique features of almost all prokaryotic cells (except for Halobacterium halobium and mycoplasmas) are cell wall peptidoglycan and the specific enzymes involved in its biosynthesis. These enzymes are target sites for inhibition of peptidoglycan synthesis by specific antibiotics. The primary chemical structures of peptidoglycans of both Gram-positive and Gram-negative bacteria have been established; they consist of a glycan backbone of repeating groups of b1, 4-linked disaccharides of b1,4-N-acetylmuramyl-N-acetylglucosamine. Tetrapeptides of L-alanine-D-isoglutamic acid-L-lysine (or diaminopimelic acid)-n-alanine are linked through the carboxyl group by amide linkage of muramic acid residues of the glycan chains; the D-alanine residues are directly cross-linked to the e-amino group of lysine or diaminopimelic acid on a neighboring tetrapeptide, or they are linked by a peptide bridge. In S aureus peptidoglycan, a glycine pentapeptide bridge links the two adjacent peptide structures. The extent of direct or peptide-bridge cross-linking varies from one peptidoglycan to another. The staphylococcal peptidoglycan is highly cross-linked, whereas that of E coli is much less so, and has a more open peptidoglycan mesh. The diamino acid providing the e-amino group for cross-linking is lysine or diaminopimelic acid, the latter being uniformly present in Gram-negative peptidoglycans. A peptidoglycan with a chemical structure substantially different from that of all eubacteria has been discovered in certain archaebacteria. Instead of muramic acid, this peptidoglycan contains talosaminuronic acid and lacks the D-amino acids found in the eubacterial peptidoglycans. Interestingly, organisms containing this wall polymer (referred to as pseudomurein) are insensitive to penicillin, an inhibitor of the transpeptidases involved in peptidoglycan biosynthesis in eubacteria.

The ß-1,4 glycosidic bond between N-acetylmuramic acid and N-acetylglucosamine is specifically cleaved by the bacteriolytic enzyme lysozyme. Widely distributed in nature, this enzyme is present in human tissues and secretions and can cause complete digestion of the peptidoglycan walls of sensitive organisms. When lysozyme is allowed to digest the cell wall of Gram-positive bacteria suspended in an osmotic stabilizer (such as sucrose), protoplasts are formed. These protoplasts are able to survive and continue to grow on suitable media in the wall-less state. Gram-negative bacteria treated similarly produce spheroplasts, which retain much of the outer membrane structure. The dependence of bacterial shape on the peptidoglycan is shown by the transformation of rod-shaped bacteria to spherical protoplasts (spheroplasts) after enzymatic breakdown of the peptidoglycan. The mechanical protection afforded by the wall peptidoglycan layer is evident in the osmotic fragility of both protoplasts and spheroplasts.

There are two groups of bacteria that lack the protective cell wall peptidoglycan structure, the Mycoplasma species, one of which causes atypical pneumonia and some genitourinary tract infections and the L-forms, which originate from Gram-positive or Gram-negative bacteria and are so designated because of their discovery and description at the Lister Institute, London. The mycoplasmas and L-forms are all Gram-negative and insensitive to penicillin and are bounded by a surface membrane structure. L-forms arising "spontaneously" in cultures or isolated from infections are structurally related to protoplasts and spheroplasts; all three forms (protoplasts, spheroplasts, and L-forms) revert infrequently and only under special conditions.

Teichoic Acids

Wall teichoic acids are found only in certain Gram-positive bacteria (such as staphylococci, streptococci, lactobacilli, and Bacillus spp); so far, they have not been found in gram- negative organisms. Teichoic acids are polyol phosphate polymers, with either ribitol or glycerol linked by phosphodiester bonds; their structures are illustrated in Figure below. Substituent groups on the polyol chains can include D-alanine (ester linked), N-acetylglucosamine, N-acetylgalactosamine, and glucose; the substituent is characteristic for the teichoic acid from a particular bacterial species and can act as a specific antigenic determinant. Teichoic acids are covalently linked to the peptidoglycan. These highly negatively charged polymers of the bacterial wall can serve as a cation-sequestering mechanism.

ACCESSORY WALL POLYMERS

In addition to the principal cell wall polymers, the walls of certain Gram-positive bacteria possess polysaccharide molecules linked to the peptidoglycan. For example, the C polysaccharide of streptococci confers group specificity. Acidic polysaccharides attached to the peptidoglycan are called teichuronic acids. Mycobacteria have peptidoglycolipids, glycolipids, and waxes associated with the cell wall.

Lipopolysaccharides

A characteristic feature of Gram-negative bacteria is possession of various types of complex macromolecular lipopolysaccharide (LPS). So far, only one Gram-positive organism, Listeria monocytogenes, has been found to contain an authentic LPS. The LPS of this bacterium and those of all Gram-negative species are also called endotoxins, thereby distinguishing these cell-bound, heat-stable toxins from heat-labile, protein exotoxins secreted into culture media. Endotoxins possess an array of powerful biologic activities and play an important role in the pathogenesis of many Gram-negative bacterial infections.

In addition to causing endotoxic shock, LPS is pyrogenic, can activate macrophages and complement, is mitogenic for B lymphocytes, induces interferon production, causes tissue necrosis and tumor regression, and has adjuvant properties. The endotoxic properties of LPS reside largely in the lipid A components. Usually, the LPS molecules have three regions: the lipid A structure required for insertion in the outer leaflet of the outer membrane bilayer; a covalently attached core composed of 2-keto-3deoxyoctonic acid (KDO), heptose, ethanolamine, N-acetylglucosamine, glucose, and galactose; and polysaccharide chains linked to the core.

The polysaccharide chains constitute the O-antigens of the Gram-negative bacteria, and the individual monosaccharide constituents confer serologic specificity on these components. Figure 2-10 depicts the structure of LPS. Although it has been known that lipid A is composed of b1,6-linked D-glucosamine disaccharide substituted with phosphomonester groups at positions 4' and 1, uncertainties have existed about the attachment positions of the six fatty acid acyl and KDO groups on the disaccharide. The demonstration of the structure of lipid A of LPS of a heptoseless mutant of Salmonella typhimurium has established that amide-linked hydroxymyristoyl and lauroxymyristoyl groups are attached to the nitrogen of the 2- and 2'-carbons, respectively, and that hydroxymyristoyl and myristoxymyristoyl groups are attached to the oxygen of the 3- and 3'-carbons of the disaccharide, respectively. Therefore, only position 6' is left for attachment of KDO units.

LPS and phospholipids help confer asymmetry to the outer membrane of the Gram-negative bacteria, with the hydrophilic polysaccharide chains outermost. Each LPS is held in the outer membrane by relatively weak cohesive forces (ionic and hydrophobic interactions) and can be dissociated from the cell surface with surface-active agents.

As in peptidoglycan biosynthesis, LPS molecules are assembled at the plasma or inner membrane. These newly formed molecules are initially inserted into the outer-inner membrane adhesion sites.

Outer Membrane of Gram-Negative Bacteria

In thin sections, the outer membranes of Gram-negative bacteria appear broadly similar to the plasma or inner membranes; however, they differ from the inner membranes and walls of Gram-positive bacteria in numerous respects. The lipid A of LPS is inserted with phospholipids to create the outer leaflet of the bilayer structure; the lipid portion of the lipoprotein and phospholipid form the inner leaflet of the outer membrane bilayer of most Gram-negative bacteria.

In addition to these components, the outer membrane possesses several major outer membrane proteins; the most abundant is called porin. The assembled subunits of porin form a channel that limits the passage of hydrophilic molecules across the outer membrane barrier to those having molecular weights that are usually less than 600 to 700.

Evidence also suggests that hydrophobic pathways exist across the outer membrane and are partly responsible for the differential penetration and effectiveness of certain b-lactam antibiotics (ampicillin, cephalosporins) that are active against various Gram-negative bacteria. Although the outer membranes act as a permeability barrier or molecular sieve, they do not appear to possess energy-transducing systems to drive active transport. Several outer membrane proteins, however, are involved in the specific uptake of metabolites (maltose, vitamin B12, nucleosides) and iron from the medium.

Thus, outer membranes of the Gram-negative bacteria provide a selective barrier to external molecules and thereby prevent the loss of metabolite-binding proteins and hydrolytic enzymes (nucleases, alkaline phosphatase) found in the periplasmic space.

The periplasmic space is the region between the outer surface of the inner (plasma) membrane and the inner surface of the outer membrane. Thus, Gram-negative bacteria have a cellular compartment that has no equivalent in Gram-positive organisms.

In addition to the hydrolytic enzymes, the periplasmic space holds binding proteins (proteins that specifically bind sugars, amino acids, and inorganic ions) involved in membrane transport and chemotactic receptor activities. Moreover, plasmid-encoded b-lactamases and aminoglycoside-modifying enzymes (phosphorylation or adenylation) in the periplasmic space produce antibiotic resistance by degrading or modifying an antibiotic in transit to its target sites on the membrane (penicillin-binding proteins) or on the ribosomes (aminoglycosides).

These periplasmic proteins can be released by subjecting the cells to osmotic shock and after treatment with the chelating agent ethylenediaminetetraacetic acid.

INTRACELLULAR COMPONENTS

Plasma (Cytoplasmic) Membranes

Bacterial plasma membranes, the functional equivalents of eukaryotic plasma membranes, are referred to variously as cytoplasmic, protoplast, or (in Gram-negative organisms) inner membranes. Similar in overall dimensions and appearance in thin sections to biomembranes from eukaryotic cells, they are composed primarily of proteins and lipids (principally phospholipids). Protein-to-lipid ratios of bacterial plasma membranes are approximately 3: 1, close to those for mitochondrial membranes. Unlike eukaryotic cell membranes, the bacterial membrane (except for Mycoplasma species and certain methylotrophic bacteria) has no sterols, and bacteria lack the enzymes required for sterol biosynthesis.

Although their composition is similar to that of inner membranes of Gram-negative species, cytoplasmic membranes from Gram-positive bacteria possess a class of macromolecules not present in the Gram-negative membranes. Many Gram-positive bacterial membranes contain membrane-bound lipoteichoic acid, and species lacking this component (such as Micrococcus and Sarcina spp) contain an analogous membrane-bound succinylated lipomannan. Lipoteichoic acids are structurally similar to the cell wall glycerol teichoic acids in that they have basal polyglycerol phosphodiester 1-3 linked chains. These chains terminate with the phosphomonoester end of the polymer, which is linked covalently to either a glycolipid or a phosphatidyl glycolipid moiety. Thus, a hydrophobic tail is provided for anchoring in the membrane lipid layers. As in the cell wall glycerol teichoic acid, the lipoteichoic acids can have glycosidic and D-alanyl ester substituents on the C-2 position of the glycerol.

Both membrane-bound lipoteichoic acid and membrane-bound succinylated lipomannan can be detected as antigens on the cell surface, and the glycerol-phosphate and succinylated mannan chains appear to extend through the cell wall structure. This class of polymer has not yet been found in the cytoplasmic membranes of Gram-negative organisms. In both instances, the lipoteichoic acids and the lipomannans are negatively charged components and can sequester positively charged substances. They have been implicated in adhesion to host cells, but their functions remain to be elucidated.

Multiple functions are performed by the plasma membranes of both Gram-positive and Gram-negative bacteria. Plasma membranes are the site of active transport, respiratory chain components, energy-transducing systems, the H+-ATPase of the proton pump, and membrane stages in the biosynthesis of phospholipids, peptidoglycan, LPS, and capsular polysaccharides. In essence, the bacterial cytoplasmic membrane is a multifunction structure that combines the mitochondrial transport and biosynthetic functions that are usually compartmentalized in discrete membranous organelles in eukaryotic cells. The plasma membrane is also the anchoring site for DNA and provides the cell with a mechanism (as yet unknown) for separation of sister chromosomes.

Mesosomes

Thin sections of Gram-positive bacteria reveal the presence of vesicular or tubular-vesicular membrane structures called mesosomes, which are apparently formed by an invagination of the plasma membrane. These structures are much more prominent in Gram-positive than in Gram-negative organisms. At one time, the mesosomal vesicles were thought to be equivalent to bacterial mitochondria; however, many other membrane functions have also been attributed to the mesosomes. At present, there is no satisfactory evidence to suggest that they have a unique biochemical or physiologic function. Indeed, electron-microscopic studies have suggested that the mescsomes, as usually seen in thin sections, may arise from membrane perturbation and fixation artifacts. No general agreement exists about this theory, however, and some evidence indicates that mesosomes may be related to events in the cell division cycle.

Other Intracellular Components

In addition to the nucleoid and cytoplasm (cytosol), the intracellular compartment of the bacterial cell is densely packed with ribosomes of the 70S type. These ribonucleoprotein particles, which have a diameter of 18 nm, are not arranged on a membranous rough endoplasmic reticulum as they are in eukaryotic cells. Other granular inclusions randomly distributed in the cytoplasm of various species include metabolic reserve particles such as poly-b-hydroxybutyrate (PHB), polysaccharide and glycogen-like granules, and polymetaphosphate or metachromatic granules.

Endospores are highly heat-resistant, dehydrated resting cells formed intracellularly in members of the genera Bacillus and Clostridium. Sporulation, the process of forming endospores, is an unusual property of certain bacteria. The series of biochemical and morphologic changes that occur during sporulation represent true differentiation within the cycle of the bacterial cell. The process, which usually begins in the stationary phase of the vegetative cell cycle, is initiated by depletion of nutrients (usually readily utilizable sources of carbon or nitrogen, or both). The cell then undergoes a highly complex, well-defined sequence of morphologic and biochemical events that ultimately lead to the formation of mature endospores.

As many as seven distinct stages have been recognized by morphologic and biochemical studies of sporulating Bacillus species: stage 0, vegetative cells with two chromosomes at the end of exponential growth; stage I, formation of axial chromatin filament and excretion of exoenzymes, including proteases; stage II, forespore septum formation and segregation of nuclear material into two compartments; stage III, spore protoplast formation and elevation of tricarboxylic acid and glyoxylate cycle enzyme levels; stage IV, cortex formation and refractile appearance of spore; stage V, spore coat protein formation; stage VI, spore maturation, modification of cortical peptidoglycan, uptake of dipicolinic acid (a unique endospore product) and calcium, and development of resistance to heat and organic solvents; and stage VII, final maturation and liberation of endospores from mother cells (in some species).

When newly formed, endospores appear as round, highly refractile cells within the vegetative cell wall, or sporangium. Some strains produce autolysins that digest the walls and liberate free endospores. The spore protoplast, or core, contains a complete nucleus, ribosomes, and energy generating components that are enclosed within a modified cytoplasmic membrane. The peptidoglycan spore wall surrounds the spore membrane; on germination, this wall becomes the vegetative cell wall. Surrounding the spore wall is a thick cortex that contains an unusual type of peptidoglycan, which is rapidly released on germination. A spore coat of keratinlike protein encases the spore contained within a membrane (the exosporium). During maturation, the spore protoplast dehydrates and the spore becomes refractile and resistant to heat, radiation, pressure, desiccation, and chemicals; these properties correlate with the cortical peptidoglycan and the presence of large amounts of calcium dipicolinate.

Recent evidence indicated that the spores of Bacillus spharicus were revived which had been preserved in amber for more than 25 million years. Their claims need to be reevaluated. The thin section of the spore shows the ruptured, thick spore coat and the cortex surrounding the spore protoplast with the germinal cell wall that becomes the vegetative wall on outgrowth.

2

Classification

INTRODUCTION

Bacteria are classified and identified to distinguish one organism from another and to group similar organisms by criteria of interest to microbiologists or other scientists. The most important level of this type of classification is the species level.

A species name should mean the same thing to everyone. Within one species, strains and subgroups can differ by the disease they produce, their environmental habitat, and many other characteristics. Formerly, species were created on the basis of such criteria, which may be extremely important for clinical microbiologists and physicians but which are not a sufficient basis for establishing a species. Verification of existing species and creation of new species should involve biochemical and other phenotypic criteria as well as DNA relatedness. In numerical or phenetic approaches to classification, strains are grouped on the basis of a large number of phenotypic characteristics. DNA relatedness is used to group strains on the basis of overall genetic similarity.

Species are identified in the clinical laboratory by morphological traits and biochemical tests, some of which are supplemented by serologic assessments (e.g., identification of *Salmonella* and *Shigella* species). Because of differences in pathogenicity *(Escherichia coli)* or the necessity to characterize a disease outbreak (*Vibrio cholerae,* methicillin-resistant *Staphylococcus aureus),* strains of medical interest are often classified below the species level by serology or identification of toxins. Pathogenic or epidemic strains also can be classified by the presence of a specific plasmid, by their plasmid profile (the number and sizes of plasmids), or by bacteriophage susceptibility patterns (phage typing). Newer molecular biologic techniques have enabled scientists to identify some species and strains (without the use of biochemical tests) by identifying a specific gene or genetic sequence, sometimes directly from the clinical specimen.

Laboratories have no difficulty in identifying typical strains of common bacteria using commonly available test systems. Problems do arise, however, when atypical strains or rare or newly described species are not in the data base. Such difficulties are compounded when the strains are misidentified rather than unidentified, and so laboratory personnel and physicians (at least infectious diseases specialists) should be familiar with taxonomic reference texts and journals that publish papers on new species. Bacterial nomenclature at the genus and species level changes often, based primarily on the use of newer genetic techniques. A species may acquire more than one name. In some cases the recognition of a new species results in a unique correlation with specific clinical problems. For example, recognition of *Porphyromonas gingivalis* as a unique species, separate from its previous inclusion within *Bacteroides melaninogenicus* (now known to be composed of several taxonomic groups of black-pigmenting anaerobic gram-negative bacilli), elucidated its role as a key pathogen in adult periodontitis. It is important to understand why these changes and synonyms exist in taxonomy.

The clinical laboratory is concerned with the rapid, sensitive, and accurate identification of microbes involved in producing disease. The number and types of tests done in such a laboratory depend on its size and the population it serves. Highly specialized or rarely performed tests should be done only by reference laboratories. Physicians, clinical laboratory personnel, and reference laboratory personnel must have a good working relationship if patients are to receive first-rate care.

In addition, the physician and the clinical laboratory personnel must know which diseases and isolates are reportable to public health laboratories and how to report them .

DEFINITIONS

Taxonomy

Taxonomy is the science of classification, identification, and nomenclature. For classification purposes, organisms are usually organized into subspecies, species, genera, families, and higher orders. For eukaryotes, the definition of the species usually stresses the ability of similar organisms to reproduce sexually with the formation of a zygote and to produce fertile offspring. However, bacteria do not undergo sexual reproduction in the eukaryotic sense. Other criteria are used for their classification.

Classification

Classification is the orderly arrangement of bacteria into groups. There is nothing inherently scientific about classification, and different groups of scientists may classify the same organisms differently. For example, clinical microbiologists are interested in the serotype, antimicrobial resistance pattern, and toxin and invasiveness factors in *Escherichia coli*, whereas geneticists are concerned with specific mutations and plasmids.

Identification

Identification is the practical use of classification criteria to distinguish certain organisms from others, to verify the authenticity or utility of a strain or a particular reaction, or to isolate and identify the organism that causes a disease.

Nomenclature

Nomenclature (naming) is the means by which the characteristics of a species are defined and communicated among microbiologists. A species name should mean the same thing to all microbiologists, yet some definitions vary in different countries or microbiologic specialty groups. For example, the organism known as *Clostridium perfringens* in the United States is called *Clostridium welchii* in England.

Species

A bacterial species is a distinct organism with certain characteristic features, or a group of organisms that resemble one another closely in the most important features of their organization. In the past, unfortunately, there was little agreement about these criteria or about the number of features necessary to distinguish a species. Species were often defined solely by such criteria as host range, pathogenicity, or ability to produce gas during the fermentation of a given sugar. Without a universal consensus, criteria reflected the interests of the investigators who described a particular species. For example, bacteria that caused plant diseases were often defined by the plant from which they were isolated; also, each new *Salmonella* serotype that was discovered was given species status. These practices have been replaced by generally accepted genetic criteria that can be used to define species in all groups of bacteria.

APPROACHES TO TAXONOMY

Numerical Approach

In their studies on members of the family Enterobacteriaceae, Edwards and Ewing established the following principles to characterize, classify, and identify organisms (Lennette et al., 1985):

Classification and identification of an organism should be based on its overall morphologic and biochemical pattern. A single characteristic (pathogenicity, host range, or biochemical reaction), regardless of its importance, is not a sufficient basis for classifying or identifying an organism.

A large and diverse strain sample must be tested to determine accurately the biochemical characteristics used to distinguish a given species.

Atypical strains often are perfectly typical members of a given biogroup within an existing species, but sometimes they are typical members of an unrecognized new species.

In numerical taxonomy (also called computer or phenetic taxonomy) many (50 to 200) biochemical, morphological, and cultural characteristics, as well as susceptibilities to antibiotics and inorganic compounds, are used to determine the degree of similarity between organisms. In numerical studies, investigators often calculate the coefficient of similarity or percentage of similarity between strains (where strain indicates a single isolate from a specimen). A dendrogram or a similarity matrix is constructed that joins individual strains into groups and places one group with other groups on the basis of their percentage of similarity. In the dendrogram in Figure 3-1, group 1 represents three *Citrobacter freundii* strains that are about 95 percent similar and join with a fourth *C freundii* strain at the level of 90 percent similarity. Group 2 is composed of three *Citrobacter diversus* strains that are 95 percent similar, and group 3 contains two *E coli* strains that are 95 percent similar, as well as a third *E coli* strain to which they are 90 percent similar. Similarity between groups 1 and 2 occurs at the 70 percent level, and group 3 is about 50 percent similar to groups 1 and 2.

In some cases, certain characteristics may be weighted more heavily; for example, the presence of spores in *Clostridium* might be weighted more heavily than the organism's ability to use a specific carbon source. A given level of similarity can be equated with relatedness at the genus, species, and, sometimes, subspecies levels. For instance, strains of a given species may cluster at a 90% similarity level, species within a given genus may cluster at the 70 percent level, and different genera in the same family may cluster at the 50 percent or lower level.

When this approach is the only basis for defining a species, it is difficult to know how many and which tests should be chosen; whether and how the tests should be weighted; and what level of similarity should be chosen to reflect relatedness at the genus and species levels.

Most bacteria have enough DNA to specify some 1,500 to 6,000 average-sized genes. Therefore, even a battery of 300 tests would assay only 5 to 20 percent of the genetic potential of a bacterium. Tests that are comparatively simple to conduct (such as those for carbohydrate utilization and for enzymes, presence of which can be assayed colorimetrically) are performed more often than tests for structural, reproductive, and regulatory genes, presence of which is difficult to assay. Thus, major differences may go undetected.

Other types of errors may occur when species are classified solely on the basis of phenotype. For example, different enzymes (specified by different genes) may catalyze the same reaction. Also, even if a metabolic gene is functional, negative reactions can occur because of the inability of the substrate to enter the cell, because of a mutation in a regulatory gene, or by production of an inactive protein.

There is not necessarily a one-to-one correlation between a reaction and the number of genes needed to carry out that reaction. For instance, six enzymatic steps may be involved in a given pathway.

If an assay for the end product is performed, a positive reaction indicates the presence of all six enzymes, whereas a negative reaction can mean the absence or nonfunction of one to six enzymes. Several other strain characteristics can affect phenotypic characterization; these include growth rate, incubation temperature, salt requirement, and pH. Plasmids that carry metabolic genes can enable strains to carry out reactions atypical for strains of that species.

The same set of "definitive" reactions cannot be used to classify all groups of organisms, and there is no standard number of specific reactions that allows identification of a species. Organisms are identified on the basis of phenotype, but, from the taxonomic standpoint, definition of species solely on this basis is subject to error.

Phylogenetic Approach

The ideal means of identifying and classifying bacteria would be to compare each gene sequence in a given strain with the gene sequences for every known species. This cannot be done, but the total DNA of one organism can be compared with that of any other organism by a method called nucleic acid hybridization or DNA hybridization. This method can be used to measure the number of DNA sequences that any two organisms have in common and to estimate the percentage of divergence within DNA sequences that are related but not identical. DNA relatedness studies have been done for yeasts, viruses, bacteriophages, and many groups of bacteria.

Five factors can be used to determine DNA relatedness: genome size, guanine-plus-cytosine (G+C) content, DNA relatedness under conditions optimal for DNA reassociation, thermal stability of related DNA sequences, and DNA relatedness under conditions supraoptimal for DNA reassociation. Because it is not practical to conduct these genotypic or phylogenetic evaluations in clinical laboratories, the results of simpler tests usually must be correlated with known phylogenetic data. For example, yellow strains of *Enterobacter cloacae* were shown, by DNA relatedness, to form a separate species, *Enterobacter sakazakii,* but were not designated as such until results of practical tests were correlated with the DNA data to allow routine laboratories to identify the new species.

Genome Size

True bacterial DNAs have genome sizes (measured as molecular weight) between 1 X 109 and 8 X 109. Genome size determinations sometimes can distinguish between groups. They were used to distinguish *Legionella pneumophila* (the legionnaire's disease bacterium) from *Bartonella (Rickettsia) quintana,* the agent of trench fever. *L pneumophila* has a genome size of about 3 X 109; that of *B quintana* is about 1 X 109.

Guanine-plus-Cytosine Content

The G+C content in bacterial DNA ranges from about 25 to 75 percent. This percentage is specific, but not exclusive, for a species; two strains with a similar G+C content may or may not belong to the same species. If the G+C contents are very different, however, the strains cannot be members of the same species.

DNA RELATEDNESS UNDER CONDITIONS OPTIMAL FOR DNA REASSOCIATION

DNA relatedness is determined by allowing single-stranded DNA from one strain to reassociate with single-stranded DNA from a second strain, to form a double-stranded DNA molecule. This is a specific, temperature-dependent reaction. The optimal temperature for DNA reassociation is 25 to 30°C below the temperature at which native double-stranded DNA denatures into single strands. Many studies indicate that a bacterial species is composed of strains that are 70 to 100 percent related. In contrast, relatedness between different species is 0 to about 65 percent. It is important to emphasize that the term "related" does not mean "identical" or "homologous." Similar but nonidentical nucleic acid sequences can reassociate.

THERMAL STABILITY OF RELATED DNA SEQUENCES

Each 1 percent of unpaired nucleotide bases in a double-stranded DNA sequence causes a 1 percent decrease in the thermal stability of that DNA duplex. Therefore, a comparison between the thermal stability of a control double-stranded molecule (in which both strands of DNA are from the same organism) and that of a heteroduplex (DNA strands from two different organisms) allows assessment of divergence between related nucleotide sequences.

DNA Relatedness under Supraoptimal Conditions for DNA Reassociation

When the incubation temperature used for DNA reassociation is raised from 25-30° C below the denaturation temperature to only 10-15° C below the denaturation temperature, only very closely related (and therefore highly thermally stable) DNA sequences can reassociate. Strains from the same species are 60 percent or more related at these supraoptimal incubation temperatures.

Defining Species on the Basis of DNA Relatedness

Use of these five factors allows a species definition based on DNA. Thus, *E coli* can be defined as a series of strains with a G+C content of 49 to 52 moles percent, a genome molecular weight of 2.3 X 109 to 3.0 X 109, relatedness of 70 percent or more at an optimal reassociation temperature with 0 to 4 percent divergence in related sequences, and relatedness of 60 percent or more at a supraoptimal reassociation temperature. Experience with more than 300 species has produced an arbitrary phylogenetic definition of a species to which most taxonomists subscribe: "strains with approximately 70% or greater DNA-DNA relatedness and with 5° C or less divergence in related sequences." When these two criteria are met, genome size and G+C content are always similar, and relatedness is almost always 60 percent or more at supraoptimal incubation temperatures. The 70 percent species relatedness rule has been ignored occasionally when the existing nomenclature is deeply ingrained, as is that for *E coli* and the four *Shigella* species. Because these organisms are all 70 percent or more related, DNA studies indicate that they should be grouped into a single species, instead of the present five species in two genera. This change has not been made because of the presumed confusion that would result.

DNA relatedness provides one species definition that can be applied equally to all organisms. Moreover, it cannot be affected by phenotypic variation, mutations, or the presence or absence of metabolic or other plasmids. It measures overall relatedness, and these factors affect only a very small percentage of the total DNA.

Polyphasic Approach

In practice, the approach to bacterial taxonomy should be polyphasic. The first step is phenotypic grouping of strains by morphological, biochemical and any other characteristics of interest. The phenotypic groups are then tested for DNA relatedness to determine whether the observed phenotypic homogeneity (or heterogeneity) is reflected by phylogenetic homogeneity or heterogeneity. The third and most important step is reexamination of the biochemical characteristics of the DNA relatedness groups. This allows determination of the biochemical borders of each group and determination of reactions of diagnostic value for the group. For identification of a given organism, the importance of specific tests is weighted on the basis of correlation with DNA results. Occasionally, the reactions commonly used will not distinguish completely between two distinct DNA relatedness groups. In these cases, other biochemical tests of diagnostic value must be sought.

PHENOTYPIC CHARACTERISTICS USEFUL IN CLASSIFICATION AND IDENTIFICATION

Morphologic Characteristics

Both wet-mounted and properly stained bacterial cell suspensions can yield a great deal of information. These simple tests can indicate the Gram reaction of the organism; whether it is acid-fast; its motility; the arrangement of its flagella; the presence of spores, capsules, and inclusion bodies; and, of course, its shape. This information often can allow identification of an organism to the genus level, or can minimize the possibility that it belongs to one or another group. Colony characteristics and pigmentation are also quite helpful. For example, colonies of several *Porphyromonas* species autofluoresce under long-wavelength ultraviolet light, and *Proteus* species swarm on appropriate media.

Growth Characteristics

A primary distinguishing characteristic is whether an organism grows aerobically, anaerobically, facultatively (i.e., in either the presence or absence of oxygen), or microaerobically (i.e., in the presence of a less than atmospheric partial pressure of oxygen). The proper atmospheric conditions are essential for isolating and identifying bacteria. Other important growth assessments include the incubation temperature, pH, nutrients required, and resistance to antibiotics. For example, one diarrheal disease agent, *Campylobacter jejuni,* grows well at 42° C in the presence of several antibiotics; another, *Y enterocolitica,* grows better than most other bacteria at 4° C. *Legionella, Haemophilus,* and some other pathogens require specific growth factors, whereas *E coli* and most other Enterobacteriaceae can grow on minimal media.

Antigens and Phage Susceptibility

Cell wall (O), flagellar (H), and capsular (K) antigens are used to aid in classifying certain organisms at the species level, to serotype strains of medically important species for epidemiologic purposes, or to identify serotypes of public health importance. Serotyping is also sometimes used to distinguish strains of exceptional virulence or public health importance, for example with *V cholerae* (O1 is the pandemic strain) and *E coli* (enterotoxigenic, enteroinvasive, enterohemorrhagic, and enteropathogenic serotypes).

Phage typing (determining the susceptibility pattern of an isolate to a set of specific bacteriophages) has been used primarily as an aid in epidemiologic surveillance of diseases caused by *Staphylococcus aureus,* mycobacteria, *P aeruginosa, V cholerae,* and *S typhi.* Susceptibility to bacteriocins has also been used as an epidemiologic strain marker. In most cases recently, phage and bacteriocin typing have been supplanted by molecular methods.

Biochemical Characteristics

Most bacteria are identified and classified largely on the basis of their reactions in a series of biochemical tests. Some tests are used routinely for many groups of bacteria (oxidase, nitrate reduction, amino acid degrading enzymes, fermentation or utilization of carbohydrates); others are restricted to a single family, genus, or species (coagulase test for staphylococci, pyrrolidonyl arylamidase test for Gram-positive cocci).

Both the number of tests needed and the actual tests used for identification vary from one group of organisms to another. Therefore, the lengths to which a laboratory should go in detecting and identifying organisms must be decided in each laboratory on the basis of its function, the type of population it serves, and its resources. Clinical laboratories today base the extent of their work on the clinical relevance of an isolate to the particular patient from which it originated, the public health significance of complete identification, and the overall cost-benefit analysis of their procedures. For example, the Centers for Disease Control and Prevention (CDC) reference laboratory uses at least 46 tests to identify members of the Enterobacteriaceae, whereas most clinical laboratories, using commercial identification kits or simple rapid tests, identify isolates with far fewer criteria.

CLASSIFICATION BELOW AND ABOVE THE SPECIES LEVEL

Below the Species Level

Particularly for epidemiological purposes, clinical microbiologists must distinguish strains with particular traits from other strains in the same species. For example, serotype O157:H7 *E coli* are identified in stool specimens because of their association with bloody diarrhea and subsequent hemolytic uremic syndrome.

Below the species level, strains are designated as groups or types on the basis of common serologic or biochemical reactions, phage or bacteriocin sensitivity, pathogenicity, or other characteristics. Many of these characteristics are already used and accepted: serotype, phage type, colicin type, biotype, bioserotype (a group of strains from the same species with common biochemical and serologic characteristics that set them apart from other members of the species), and pathotype (e.g., toxigenic *Clostridium difficile,* invasive *E coli,* and toxigenic *Corynebacterium diphtheriae*).

Above the Species Level

In addition to species and subspecies designations, clinical microbiologists must be familiar with genera and families. A genus is a group of related species, and a family is a group of related genera.

An ideal genus would be composed of species with similar phenotypic and phylogenetic characteristics. Some phenotypically homogeneous genera approach this criterion *(Citrobacter, Yersinia,* and *Serratia).* More often, however, the phenotypic similarity is present, but the genetic relatedness is not. *Bacillus, Clostridium,* and *Legionella* are examples of accepted phenotypic genera in which genetic relatedness between species is not 50 to 65 percent, but 0 to 65 percent. When phenotypic and genetic similarity are not both present, phenotypic similarity generally should be given priority in establishing genera. Identification practices are simplified by having the most phenotypically similar species in the same genus. The primary consideration for a genus is that it contain biochemically similar species that are convenient or important to consider as a group separate from other groups of organisms.

The sequencing of ribosomal RNA (rRNA) genes, which have been highly conserved through evolution, allows phylogenetic comparisons to be made between species whose total DNAs are essentially unrelated. It also allows phylogenetic classification at the genus, family, and higher taxonomic levels. The rRNA sequence data are usually not used to designate genera or families unless supported by similarities in phenotypic tests.

Designation of New Species and Nomenclatural Changes

Species are named according to principles and rules of nomenclature set forth in the Bacteriological Code. Scientific names are taken from Latin or Greek. The correct name of a species or higher taxon is determined by three criteria: valid publication, legitimacy of the name with regard to the rules of nomenclature, and priority of publication (that is, it must be the first validly published name for the taxon).

To be published validly, a new species proposal must contain the species name, a description of the species, and the designation of a type strain for the species, and the name must be published in the *International Journal for Systematic Bacteriology (IJSB)*. Once proposed, a name does not go through a formal process to be accepted officially; in fact, the opposite is truea validly published name is assumed to be correct unless and until it is challenged officially. A challenge is initiated by publishing a request for an opinion (to the Judicial Commission of the International Association of Microbiological Societies) in the *IJSB*. This occurs only in cases in which the validity of a name is questioned with respect to compliance with the rules of the Bacteriological Code. A question of classification that is based on scientific data (for example, whether a species, on the basis of its biochemical or genetic characteristics, or both, should be placed in a new genus or an existing genus) is not settled by the Judicial Commission, but by the preference and usage of the scientific community. This is why there are pairs of names such as *Providencia rettgeri/Proteus rettgeri, Moraxella catarrhalis/Branhamella catarrhalis,* and *Legionella micdadei/ Tatlockia micdadei.* More than one name may thus exist for a single organism. This is not, however, restricted to bacterial nomenclature. Multiple names exist for many antibiotics and other drugs and enzymes.

A number of genera have been divided into additional genera and species have been moved to new or existing genera, such as *Arcobacter* (new genus for former members of *Campylobacter*) and *Burkholderia* species (formerly species of *Pseudomonas*). Two former *Campylobacter* species (*cinaedi* and *fennelliae*) have been moved to the existing genus *Helicobacter* in another example.

The best source of information for new species proposals and nomenclatural changes is the *IJSB.* In addition, the *Journal of Clinical Microbiology* often publishes descriptions of newly described microorganisms isolated from clinical sources. Information, including biochemical reactions and sources of isolation, about new organisms of clinical importance, disease outbreaks caused by newer species, and reviews of clinical significance of certain organisms may be found in the *Annals of Internal Medicine, Journal of Infectious Diseases, Clinical Microbiology Reviews,* and *Clinical Infectious Diseases.* The data provided in these publications supplement and update *Bergey's Manual of Systematic Bacteriology,* the definitive taxonomic reference text.

Assessing Newly Described Bacteria

Since 1974, the number of genera in the family Enterobacteriaceae has increased from 12 to 28 and the number of species from 42 to more than 140, some of which have not yet been named. Similar explosions have occurred in other genera. In 1974, five species were listed in the genus *Vibrio* and four in *Campylobacter;* the genus *Legionella* was unknown. Today, there are at least 25 species in *Vibrio,* 12 *Campylobacter* species, and more than 40 species in *Legionella*. The total numbers of genera and species continue to increase dramatically.

The clinical significance of the agent of legionnaire's disease was well known long before it was isolated, characterized, and classified as *Legionella pneumophila*. In most cases, little is known about the clinical significance of a new species at the time it is first described. Assessments of clinical significance begin after clinical laboratories adopt the procedures needed to detect and identify the species and accumulate a body of data.

In fact, the detection and even the identification of uncultivatable microbes from different environments are now possible using standard molecular methods. The agents of cat scratch disease (*Bartonella henselae*) and Whipple's disease (*Tropheryma whippelii*) were elucidated in this manner. *Bartonella henselae* has since been cultured from several body sites from numerous patients; *T whippelii* remains uncultivated.

New species will continue to be described. Many will be able to infect humans and cause disease, especially in those individuals who are immunocompromised, burned, postsurgical, geriatric, and suffering from acquired immunodeficiency syndrome (AIDS). With today's severely immunocompromised patients, often the beneficiaries of advanced medical interventions, the concept of "pathogen" holds little meaning. Any organism is capable of causing disease in such patients under the appropriate conditions.

Role of the Clinical Laboratory

Clinical laboratory scientists should be able to isolate,. identify, and determine the antimicrobial susceptibility pattern of the vast majority of human disease agents so that physicians can initiate appropriate treatment as soon as possible, and the source and means of transmission of outbreaks can be ascertained to control the disease and prevent its recurrence. The need to identify clinically relevant microorganisms both quickly and cost-effectively presents a considerable challenge.

To be effective, the professional clinical laboratory staff must interact with the infectious diseases staff. Laboratory scientists should attend infectious disease rounds. They must keep abreast of new technology, equipment, and classification and should communicate this information to their medical colleagues. They should interpret, qualify, or explain laboratory reports. If a bacterial name is changed or a new species reported, the laboratory should provide background information, including a reference.

The clinical laboratory must be efficient. A concerted effort must be made to eliminate or minimize inappropriate and contaminated specimens and the performance of procedures with little or no clinical relevance. Standards for the selection, collection, and transport of specimens should be developed for both laboratory and nursing procedure manuals and reviewed periodically by a committee composed of medical, nursing, and laboratory staff. Ongoing dialogues and continuous communication with other health care workers concerning topics such as specimen collection, test selection, results interpretation, and new technology are essential to maintaining high quality microbiological services.

Biochemical and Susceptibility Testing

Most laboratories today use either commercially available miniaturized biochemical test systems or automated instruments for biochemical tests and for susceptibility testing.

The kits usually contain 10 to 20 tests. The test results are converted to numerical biochemical profiles that are identified by using a codebook or a computer. Carbon source utilization systems with up to 95 tests are also available. Most identification takes 4 to 24 hours. Biochemical and enzymatic test systems for which data bases have not been developed are used by some reference laboratories.

Automated instruments can be used to identify most Gram-negative fermenters, nonfermenters, and Gram-positive bacteria, but not for anaerobes. Antimicrobial susceptibility testing can be performed for some microorganisms with this equipment, with results expressed as approximate minimum inhibitory drug concentrations. Both tasks take 4 to 24 hours. If semiautomated instruments are used, some manipulation is done manually, and the cultures (in miniature cards or microdilution plates) are incubated outside of the instrument. The test containers are then read rapidly by the instrument, and the results are generated automatically. Instruments are also available for identification of bacteria by cell wall fatty acid profiles generated with gas-liquid chromatography (GLC), analysis of mycolic acids using high performance liquid chromatography (HPLC), and by protein-banding patterns generated by polyacrylamide gel electrophoresis (PAGE). Some other instruments designed to speed laboratory diagnosis of bacteria are those that detect (but do not identify) bacteria in blood cultures, usually faster than manual systems because of continuous monitoring. Also available are many rapid screening systems for detecting one or a series of specific bacteria, including certain streptococci, *N meningitidis,* salmonellae, *Chlamydia trachomatis*, and many others. These screening systems are based on fluorescent antibody, agglutination, or other rapid procedures.

It is important to inform physicians as soon as a presumptive identification of an etiologic agent is obtained so that appropriate therapy can be initiated as quickly as possible. Gram stain and colony morphology; acid-fast stains; and spot indole, oxidase, and other rapid enzymatic tests may allow presumptive identification of an isolate within minutes.

ROLE OF THE REFERENCE LABORATORY

Despite recent advances, the armamentarium of the clinical laboratory is far from complete. Few laboratories can or should conduct the specialized tests that are often essential to distinguish virulent from avirulent strains. Serotyping is done only for a few species, and phage typing only rarely. Few pathogenicity tests are performed. Not many laboratories can conduct comprehensive biochemical tests on strains that cannot be identified readily by commercially available biochemical systems.

Even fewer laboratories are equipped to perform plasmid profiles, gene probes, or DNA hybridization. These and other specialized tests for the serologic or biochemical identification of some exotic bacteria, yeasts, molds, protozoans, and viruses are best done in regional reference laboratories.

It is not cost-effective for smaller laboratories to store and control the quality of reagents and media for tests that are seldom run or quite complex In addition, it is impossible to maintain proficiency when tests are performed rarely. Sensitive methods for the epidemiologic subtyping of isolates from disease outbreaks, such as electrophoretic enzyme typing, rRNA fingerprinting, whole-cell protein electrophoretic patterns, and restriction endonuclease analysis of whole-cell or plasmid DNA, are used only in reference laboratories and a few large medical centers.

Specific genetic probes are now available commercially for identifying virulence factors and many bacteria and viruses. Genetic probes are among the most common methods used for identification of *Mycobacterium tuberculosis* and *M avium* complex in the U.S. today. Probes for *Neisseria gonorrhoeae* and *Chlamydia trachomatis* are now being used directly on clinical specimens with excellent sensitivity and almost universal specificity with same-day results. Mycobacterial probes are also being evaluated for direct specimen testing.

Interfacing with Public Health Laboratories

Hospital and local clinical laboratories interact with district, state, and federal public health laboratories in several important ways. The clinical laboratories participate in quality control and proficiency testing programs that are conducted by federally regulated agencies. The government reference laboratories supply cultures and often reagents for use in quality control, and they conduct training programs for clinical laboratory personnel.

All types of laboratories should interact closely to provide diagnostic services and epidemic surveillance. The primary concern of the clinical laboratory is identifying infectious disease agents and studying nosocomial and local outbreaks of disease. When the situation warrants, the local laboratory may ask the state laboratory for help in identifying an unusual organism, discovering the cause or mode of transmission in a disease outbreak, or performing specialized tests not done routinely in clinical laboratories.

Cultures should be pure and should be sent on appropriate media following appropriate procedures for transport of biohazardous materials. Pertinent information, including the type of specimen; patient name (or number), date of birth, and sex; clinical diagnosis, associated illness, date of onset, and present condition; specific agent suspected, and any other organisms isolated; relevant epidemiologic and clinical data; treatment of patient; previous laboratory results (biochemical or serologic tests); and necessary information about the submitting party must accompany each request.

These data allow the state laboratory to test the specimen properly and quickly, and they provide information about occurrences within the state. For example, a food-borne outbreak might extend to many parts of the state (or beyond its boundaries). The state laboratory can alert local physicians to the possibility of such outbreaks.

Another necessary interaction between local and state laboratories is the reporting of notifiable diseases by the local laboratory. The state laboratory makes available to local laboratories summaries of the incidence of these diseases. The state laboratories also submit the summaries to the CDC weekly (or, for some diseases, yearly), and national summaries are published weekly in the *Morbidity and Mortality Weekly Report*.

Interaction between the CDC and state and federal laboratories is very similar to that between local and state laboratories.

The CDC provides quality control cultures and reagents to state laboratories, and serves as a national reference laboratory for diagnostic services and epidemiologic surveillance. Local laboratories, however, must initially send specimens to the local or state public health laboratory, which, when necessary, forwards them to the CDC. The CDC reports its results back to the state laboratory, which then reports to the local laboratory.

HAZARDS OF CLINICAL LABORATORY WORK

Clinical laboratory personnel, including support and clerical employees, are subject to the risk of infection, chemical hazards, and, in some laboratories, radioactive contamination. Such risks can be prevented or minimized by a laboratory safety program.

Radiation Hazards

Personnel who work with radioactive materials should have taken a radioactivity safety course; they should wear radiation monitor badges and be aware of the methods for decontaminating hands, clothing, work surfaces, and equipment. They should wear gloves when working with radioactive compounds. When they work with high-level radiation, they should use a hood and stand behind a radiation shield. Preparative radioactive work should be done in a separate room with access only by personnel who are involved directly in the work.

Chemical Hazards

Chemicals can harm laboratory personnel through inhalation or skin absorption of volatile compounds; bodily contact with carcinogens, acids, bases, and other harmful chemicals; or introduction of poisonous or skin-damaging liquids into the mouth.

Good laboratory practices require that volatile compounds be handled only under a hood, that hazardous chemicals never be pipetted by mouth, and that anyone working with skin-damaging chemicals wear gloves, eye guards, and other personal protective equipment as necessary.

Workers should be familiar with the materials safety data sheets (MSDS) posted in an accessible place in every laboratory. These forms contain information about chemical hazards and procedures for decontamination should an accident occur.

Biologic Hazards

Microbiologic contamination is the greatest hazard in clinical microbiology laboratories. Laboratory infections are a danger not only to the clinical laboratory personnel but also to anyone else who enters the laboratory, including janitors, clerical and maintenance personnel, and visitors.

The risk of infection is governed by the frequency and length of contact with the infectious agent, its virulence, the dose and route of administration, and the susceptibility of the host. The inherent hazard of any infectious agent is affected by factors such as the volume of infectious material used, handling of the material, effectiveness of safety containment equipment, and soundness of laboratory methods. Body fluids from patients, particularly those containing blood, are considered potentially infectious for blood-borne pathogens, and must be handled appropriately.

If possible, agents that are treated differently, such as viruses as opposed to bacteria, or *M tuberculosis* in contrast to *E coli,* should be handled in different laboratories or in different parts of the same laboratory. When the risk category of an agent is known, it should be handled in an area with appropriate containment.

All specimens sent for microbiological studies and all organisms sent to the laboratory for identification should be assumed to be potentially infectious. A separate area should be set aside for the receipt of specimens. Personnel should be aware of the potential hazards of improperly packed, broken, or leaking packages and of the proper methods for their handling and decontamination .

To prevent infection, personnel should wear moisture-proof laboratory coats at all times, wash their hands before and after wearing gloves and at the conclusion of each potential exposure to etiologic agents, refrain from mouth pipetting, and not eat, drink, smoke, or apply cosmetics in the laboratory. Immunization may be appropriate for employees who are exposed often to certain infectious agents, including hepatitis

B, yellow fever, rabies, polioviruses, meningococci, *Y pestis, S typhi,* and *Francisella tularensis.* Universal precautions, body substance isolation, and other mandated practices involve the use of personal protective equipment and engineering controls to minimize laboratory scientists' exposure to blood-borne pathogens, even when the risk of infection is unknown.

Biosafety Levels

Infectious agents are assigned to a biosafety level from 1 to 4 on the basis of their virulence. The containment levels for organisms should correlate with the biosafety level assigned. Biosafety level 1 is for well-defined organisms not known to cause disease in healthy humans; it includes certain nonvirulent *E coli* strains (such as K-12) and *B subtilis.* Containment level 1 involves standard microbiologic practices, and safety equipment is not needed.

Biosafety level 2, the minimum level for clinical laboratories, is for moderate-risk agents associated with human disease. Containment level 2 includes limited access to the work area, decontamination of all infectious wastes, use of protective gloves, and a biologic safety cabinet for use in procedures that may create aerosols.

Examples of biosafety level 2 agents include nematode, protozoan, trematode, and cestode human parasites; all human fungal pathogens except *Coccidioides immitis;* all members of the Enterobacteriaceae except *Y pestis; Bacillus anthracis; Clostridium tetani; Corynebacterium diphtheriae; Haemophilus* species; leptospires; legionellae; mycobacteria other than *M tuberculosis;* pathogenic *Neisseria* species; staphylococci, streptococci, *Treponema pallidum; V cholerae;* and hepatitis and influenza viruses. Clinical specimens potentially containing some biosafety level 3 agents, such as *Brucella* spp., are usually handled using biosafety level 2 containment practices.

Biosafety level 3 is for agents that are associated with risk of serious or fatal aerosol infection. In containment level 3, laboratory access is controlled, special clothing is worn in the laboratory, and containment equipment is used for all work with the agent. *M tuberculosis, Coccidioides immitis, Coxiella burnetii,* and many of the arboviruses are biosafety 3 level agents. Containment level 3 usually is recommended for work with cultures of rickettsiae, brucellae, *Y pestis,* and a wide variety of viruses, including human immunodeficiency viruses.

Biosafety level 4 indicates dangerous and novel agents that cause diseases with high fatality rates. Maximum containment and decontamination procedures are used in containment level 4, which is found in only a few reference and research laboratories. Only a few viruses (including Lassa, Ebola, and Marburg viruses) are classified in biosafety level 4.

3

Bacterial Metabolism

INTRODUCTION

Metabolism refers to all the biochemical reactions that occur in a cell or organism. The study of bacterial metabolism focuses on the chemical diversity of substrate oxidations and dissimilation reactions (reactions by which substrate molecules are broken down), which normally function in bacteria to generate energy. Also within the scope of bacterial metabolism is the study of the uptake and utilization of the inorganic or organic compounds required for growth and maintenance of a cellular steady state (assimilation reactions). These respective exergonic (energy-yielding) and endergonic (energy-requiring) reactions are catalyzed within the living bacterial cell by integrated enzyme systems, the end result being self-replication of the cell. The capability of microbial cells to live, function, and replicate in an appropriate chemical milieu (such as a bacterial culture medium) and the chemical changes that result during this transformation constitute the scope of bacterial metabolism.

The bacterial cell is a highly specialized energy transformer. Chemical energy generated by substrate oxidations is conserved by formation of high-energy compounds such as adenosine diphosphate (ADP) and adenosine triphosphate (ATP) or compounds containing the thioester bond

$$(R-\overset{\overset{\displaystyle O}{\|}}{C}\sim S-R)$$, such as acetyl ~ *S*-coenzyme A

(acetyl ~ SCoA) or succinyl ~ SCoA. ADP and ATP represent adenosine monophosphate (AMP) plus one and two high-energy phosphates (AMP ~ P and AMP ~ P~ P, respectively); the energy is stored in these compounds as high-energy phosphate bonds. In the presence of proper enzyme systems, these compounds can be used as energy sources to synthesize the new complex organic compounds needed by the cell. All living cells must maintain steady-state biochemical reactions for the formation and use of such high-energy compounds.

Kluyver and Donker (1924 to 1926) recognized that bacterial cells, regardless of species, were in many respects similar chemically to all other living cells. For example, these investigators recognized that hydrogen transfer is a common and fundamental feature of all metabolic processes. Bacteria, like mammalian and plant cells, use ATP or the high-energy phosphate bond (~ P) as the primary chemical energy source. Bacteria also require the B-complex vitamins as functional coenzymes for many oxidation-reduction reactions needed for growth and energy transformation. An organism such as *Thiobacillus thiooxidans*, grown in a medium containing only sulfur and inorganic salts, synthesizes large amounts of thiamine, riboflavine, nicotinic acid, pantothenic acid, pyridoxine, and biotin. Therefore, Kluyver proposed the unity theory of biochemistry (*Die Einheit in der Biochemie*), which states that all basic enzymatic reactions which support and maintain life processes within cells of organisms, had more similarities than differences. This concept of biochemical unity stimulated many investigators to use bacteria as model systems for studying related eukaryotic, plant and animal biochemical reactions that are essentially "identical" at the molecular level.

From a nutritional, or metabolic, viewpoint, three major physiologic types of bacteria exist: the heterotrophs (or chemoorganotrophs), the autotrophs (or chemolithotrophs), and the photosynthetic bacteria (or phototrophs). These are discussed below.

Heterotrophic Metabolism

Heterotrophic bacteria, which include all pathogens, obtain energy from oxidation of organic compounds. Carbohydrates (particularly glucose), lipids, and protein are the most commonly oxidized compounds. Biologic oxidation of these organic compounds by bacteria results in synthesis of ATP as the chemical energy source. This process also permits generation of simpler organic compounds (precursor molecules) needed by the bacteria cell for biosynthetic or assimilatory reactions.

The Krebs cycle intermediate compounds serve as precursor molecules (building blocks) for the energy-requiring biosynthesis of complex organic compounds in bacteria. Degradation reactions that simultaneously produce energy and generate precursor molecules for the biosynthesis of new cellular constituents are called amphibolic.

All heterotrophic bacteria require preformed organic compounds. These carbon- and nitrogen-containing compounds are growth substrates, which are used aerobically or anaerobically to generate reducing equivalents (e.g., reduced nicotinamide adenine dinucleotide; NADH + H^+); these reducing equivalents in turn are chemical energy sources for all biologic oxidative and fermentative systems. Heterotrophs are the most commonly studied bacteria; they grow readily in media containing carbohydrates, proteins, or other complex nutrients such as blood. Also, growth media may be enriched by the addition of other naturally occurring compounds such as milk (to study lactic acid bacteria) or hydrocarbons (to study hydrocarbon-oxidizing organisms).

Respiration

Glucose is the most common substrate used for studying heterotrophic metabolism. Most aerobic organisms oxidize glucose completely by the following reaction equation:

$$C_6H_{12}6O_2 \rightarrow 6CO_2 + 6H_2O + \text{energy}$$

This equation expresses the cellular oxidation process called respiration. Respiration occurs within the cells of plants and animals, normally generating 38 ATP molecules (as energy) from the oxidation of 1 molecule of glucose. This yields approximately 380,000 calories (cal) per mode of glucose (ATP ~ 10,000 cal/mole). Thermodynamically, the complete oxidation of one mole of glucose should yield approximately

688,000 cal; the energy that is not conserved biologically as chemical energy (or ATP formation) is liberated as heat (308,000 cal). Thus, the cellular respiratory process is at best about 55% efficient.

Glucose oxidation is the most commonly studied dissimilatory reaction leading to energy production or ATP synthesis. The complete oxidation of glucose may involve three fundamental biochemical pathways. The first is the glycolytic or Embden- Meyerhof-Parnas pathway, the second is the Krebs cycle (also called the citric acid cycle or tricarboxylic acid cycle), and the third is the series of membrane-bound electron transport oxidations coupled to oxidative phosphorylation.

Respiration takes place when any organic compound (usually carbohydrate) is oxidized completely to CO_2 and H_2O. In aerobic respiration, molecular 0_2 serves as the terminal acceptor of electrons. For anaerobic respiration, NO_3-, SO_4^{2-}, CO_2, or fumarate can serve as terminal electron acceptors (rather than 0_2), depending on the bacterium studied. The end result of the respiratory process is the complete oxidation of the organic substrate molecule, and the end products formed are primarily CO_2 and H_2O. Ammonia is formed also if protein (or amino acid) is the substrate oxidized.

Metabolically, bacteria are unlike cyanobacteria (blue-green algae) and eukaryotes in that glucose oxidation may occur by more than one pathway. In bacteria, glycolysis represents one of several pathways by which bacteria can catabolically attack glucose. The glycolytic pathway is most commonly associated with anaerobic or fermentative metabolism in bacteria and yeasts. In bacteria, other minor heterofermentative pathways, such as the phosphoketolase pathway, also exist.

In addition, two other glucose-catabolizing pathways are found in bacteria: the oxidative pentose phosphate pathway (hexose monophosphate shunt), (Fig. 4-3) and the Entner-Doudoroff pathway, which is almost exclusively found in obligate aerobic bacteria (Fig. 4-4). The highly oxidative *Azotobacter* and most *Pseudomonas* species, for example, utilize the Entner-Doudoroff pathway for glucose catabolism, because these organisms lack the enzyme phosphofructokinase and hence cannot synthesize fructose 1,6-diphosphate, a key intermediate compound in the glycolytic pathway. (Phospho-fructokinase is also sensitive to molecular 02 and does not function in obligate aerobes).

Other bacteria, which lack aldolase (which splits fructose-1,6-diphosphate into two triose phosphate compounds), also cannot have a functional glycolytic pathway. Although the Entner-Doudoroff pathway is usually associated with obligate aerobic bacteria, it is present in the facultative anaerobe *Zymomonas mobilis* (formerly *Pseudomonas lindneri*). This organism dissimilates glucose to ethanol and represents a major alcoholic fermentation reaction in a bacterium.

Glucose dissimilation also occurs by the hexose monophosphate shunt (Fig. 4-3). This oxidative pathway was discovered in tissues that actively metabolize glucose in the presence of two glycolytic pathway inhibitors (iodoacetate and fluoride). Neither inhibitor had an effect on glucose dissimilation, and NADPH + H+ generation occurred directly from the oxidation of glucose-6-phosphate (to 6-phosphoglucono-d-lactone) by glucose-6phosphate dehydrogenase. The pentose phosphate pathway subsequently permits the direct oxidative decarboxylation of glucose to pentoses. The capability of this oxidative metabolic system to bypass glycolysis explains the term shunt.

The biochemical reactions of the Entner-Doudoroff pathway are a modification of the hexose monophosphate shunt, except that pentose sugars are not directly formed. The two pathways are identical up to the formation of 6-phosphogluconate (see Fig. 4-4) and then diverge. In the Entner-Doudoroff pathway, no oxidative decarboxylation of 6-phosphogluconate occurs and no pentose compound is formed. For this pathway, a new 6 carbon compound intermediate (2-keto-3-deoxy6-phosphogluconate) is generated by the action of 6-phosphogluconate dehydratase (an Fe2+- and glutathione-stimulated enzyme); this intermediate compound is then directly cleaved into the triose (pyruvate) and a triose-phosphate compound (glyceraldehyde-3-phosphate) by the 2-keto-3-deoxy6-phosphogluconate aldolase. The glyceraldehyde-3-phosphate is further oxidized to another pyruvate molecule by the same enzyme systems that catalyze the terminal glycolytic pathway.

The glycolytic pathway may be the major one existing concomitantly with the minor oxidative pentose phosphate - hexose monophosphate shunt pathway; the Entner-Doudoroff pathway also may function as a major pathway with a minor hexose monophosphate shunt. A few bacteria possess only one pathway. All cyanobacteria, *Acetobacter suboxydans*, and *A xylinum* possess only the hexose monophosphate shunt pathway; *Pseudomonas saccharophilia* and *Z mobilis* possess solely the Entner-Doudoroff pathway.

All major pathways of glucose or hexose catabolism have several metabolic features in common. First, there are the preparatory steps by which key intermediate compounds such as the triose-PO_4, glyceraldehyde-3-phosphate, and/or pyruvate are generated. The latter two compounds are almost universally required for further assimilatory or dissimilatory reactions within the cell. Second, the major source of phosphate for all reactions involving phosphorylation of glucose or other hexoses is ATP, not inorganic phosphate (Pi). Actually, chemical energy contained in ATP must be initially spent in the first step of glucose metabolism (via kinase-type enzymes) to generate glucose-6-phosphate, which initiates the reactions involving hexose catabolism. Third, NADH + H^+ or NADPH + H^+ is generated as reducing equivalents (potential energy) directly by one or more of the enzymatic reactions involved in each of these pathways.

Fermentation

Fermentation, another example of heterotrophic metabolism, requires an organic compound as a terminal electron (or hydrogen) acceptor. In fermentations, simple organic end products are formed from the anaerobic dissimilation of glucose (or some other compound). Energy (ATP) is generated through the dehydrogenation reactions that occur as glucose is broken down enzymatically. The simple organic end products formed from this incomplete biologic oxidation process also serve as final electron and hydrogen acceptors. On reduction, these organic end products are secreted into the medium as waste metabolites (usually alcohol or acid). The organic substrate compounds are incompletely oxidized by bacteria, yet yield sufficient energy for microbial growth. Glucose is the most common hexose used to study fermentation reactions.

In the late 1850s, Pasteur demonstrated that fermentation is a vital process associated with the growth of specific microorganisms, and that each type of fermentation can be defined by the principal organic end product formed (lactic acid, ethanol, acetic acid, or butyric acid). His studies on butyric acid fermentation led directly to the discovery of anaerobic microorganisms. Pasteur concluded that oxygen inhibited the microorganisms responsible for butyric acid fermentation because both bacterial mobility and butyric acid formation ceased when air was bubbled into the fermentation mixture. Pasteur also introduced the terms aerobic and anaerobic. His views on fermentation are made clear from his microbiologic studies on the production of beer (from *Etudes sur la Biere*, 1876):

> In the experiments which we have described, fermentation by yeast is seen to be the direct consequence of the processes of nutrition, assimilation and life, when these are carried on without the agency of free oxygen. The heat required in the accomplishment of that work must necessarily have been borrowed from the decomposition of the fermentation matter.... Fermentation by yeast appears, therefore, to be essentially connected with the property possessed by this minute cellular plant of performing its respiratory functions, somehow or other, with the oxygen existing combined in sugar.

For most microbial fermentations, glucose dissimilation occurs through the glycolytic pathway. The simple organic compound most commonly generated is pyruvate, or a compound derived enzymatically from pyruvate, such as acetaldehyde, a-acetolactate, acetyl ~ SCoA, or lactyl ~ SCoA. Acetaldehyde can then be reduced by NADH + H+ to ethanol, which is excreted by the cell. The end product of lactic acid fermentation, which occurs in streptococci (e.g., *Streptococcus lactis*) and many lactobacilli (e.g., *Lactobacillus casei, L pentosus*), is a single organic acid, lactic acid. Organisms that produce only lactic acid from glucose fermentation are homofermenters. Homofermentative lactic acid bacteria dissimilate glucose exclusively through the glycolytic pathway. Organisms that ferment glucose to multiple end products, such as acetic acid, ethanol, formic acid, and CO_2, are referred to as heterofermenters. Examples of heterofermentative bacteria include *Lactobacillus, Leuconostoc,* and *Microbacterium* species. Heterofermentative fermentations are more common among bacteria, as in the mixed-acid fermentations carried out by bacteria of the family Enterobacteriaceae (e.g., *Escherichia coli, Salmonella, Shigella,* and *Proteus* species).

Many of these glucose fermenters usually produce CO_2 and H_2 with different combinations of acid end products (formate, acetate, lactate, and succinate). Other bacteria such as *Enterobacter aerogenes, Aeromonas, Serratia, Erwinia,* and *Bacillus* species also form CO_2 and H_2 as well as other neutral end products (ethanol, acetylmethylcarbinol [acetoin], and 2,3-butylene glycol). Many obligately anaerobic clostridia (e.g., *Clostridium saccharobutyricum, C thermosaccharolyticum*) and *Butyribacterium* species ferment glucose with the production of butyrate, acetate, CO_2, and H_2, whereas other *Clostridum* species (*C acetobutylicum* and *C butyricum*) also form these fermentation end products plus others (butanol, acetone, isopropanol, formate, and ethanol). Similarly, the anaerobic propionic acid bacteria (*Propionibacterium* species) and the related *Veillonella* species ferment glucose to form CO_2, propionate, acetate, and succinate. In these bacteria, propionate is formed by the partial reversal of the Krebs cycle reactions and involves a CO_2 fixation by pyruvate (the Wood-Werkman reaction) that forms oxaloacetate (a four-carbon intermediate).

Oxaloacetate is then reduced to malate, fumarate, and succinate, which is decarboxylated to propionate. Propionate is also formed by another three-carbon pathway in *C propionicum, Bacteroides ruminicola,* and *Peptostreptococcus* species, involving a lactyl ~ SCoA intermediate. The obligately aerobic acetic acid bacteria (*Acetobacter* and the related *Gluconobacter* species) can also ferment glucose, producing acetate and gluconate. For thermodynamic reasons, bacteria that rely on fermentative process for growth cannot generate as much energy as respiring cells. In respiration, 38 ATP molecules (or approximately 380,000 cal/mole) can be generated as biologically useful energy from the complete oxidation of 1 molecule of glucose (assuming 1 NAD(P)H = 3 ATP and 1 ATP ADP + Pi = 10,000 cal/mole).

Although only 2 ATP molecules are generated by this glycolytic pathway, this is apparently enough energy to permit anaerobic growth of lactic acid bacteria and the ethanolic fermenting yeast, *Saccharomyces cerevisiae*. The ATP-synthesizing reactions in the glycolytic pathway specifically involve the substrate phosphorylation reactions catalyzed by phosphoglycerokinase and pyruvic kinase. Although all the ATP molecules available for fermentative growth are believed to be generated by these substrate phosphorylation reactions, some energy equivalents are also generated by proton extrusion reactions (acid liberation), which occur with intact membrane systems and involve the proton extrusion reactions of energy conservation as it applies to fermentative metabolism.

Krebs Cycle

The Krebs cycle (also called the tricarboxylic acid cycle or citic acid cycle) functions oxidatively in respiration and is the metabolic process by which pyruvate or acetyl ~ SCoA is completely decarboxylated to CO_2. In bacteria, this reaction occurs through acetyl ~ SCoA, which is the first product in the oxidative decarboxylation of pyruvate by pyruvate dehydrogenase. If 2 pyruvate molecules are obtained from the dissimilation of 1 glucose molecule, then 30 ATP molecules are generated in total. The decarboxylation of pyruvate, isocitrate, and a-ketoglutarate accounts for all CO_2 molecules generated during the respiratory process. The chemical energy conserved by the Krebs cycle is contained in the reduced compounds is not available as ATP until the final step of respiration (electron transport and oxidative phosphorylation) occurs.

The Krebs cycle is therefore another preparatory stage in the respiratory process. If 1 molecule of pyruvate is oxidized completely to 3 molecules of CO_2, generating 15 ATP molecules, the oxidation of 1 molecule of glucose will yield as many as 38 ATP molecules, provided glucose is dissimilated by glycolysis and the Krebs cycle (further assuming that the electron transport/oxidative phosphorylation reactions are bioenergetically identical to those of eukaryotic mitochondria).

Glyoxylate Cycle

In general, the Krebs cycle functions similarly in bacteria and eukaryotic systems, but major differences are found among bacteria. One difference is that in obligate aerobes, L-malate may be oxidized directly by molecular 0_2 via an electron transport chain. In other bacteria, only some Krebs cycle intermediate reactions occur because a-ketoglutarate dehydrogenase is missing.

A modification of the Krebs cycle, commonly called the glyoxylate cycle, or shunt, which exists in some bacteria. This shunt functions similarly to the Krebs cycle but lacks many of the Krebs cycle enzyme reactions. The glyoxylate cycle is primarily an oxidative pathway in which acetyl~SCoA is generated from the oxidation, of acetate, which usually is derived from the oxidation of fatty acids. The oxidation of fatty acids to acetyl~SCoA is carried out by the b-oxidation pathway. Pyruvate oxidation is not directly involved in the glyoxylate shunt, yet this shunt yields sufficient succinate and malate, which are required for energy production. The glyoxylate cycle also generates other precursor compounds needed for biosynthesis.

The glyoxylate cycle was discovered as an unusual metabolic pathway during an attempt to learn how lipid (or acetate) oxidation in bacteria and plant seeds could lead to the direct biosynthesis of carbohydrates. The glyoxylate cycle converts oxaloacetate either to pyruvate and CO_2 (catalyzed by pyruvate carboxylase) or to phosphoenolpyruvate and CO_2 (catalyzed by the inosine triphosphate [ITP]-dependent phosphoenolpyruvate carboxylase kinase). Either triose compound can then be converted to glucose by reversal of the glycolytic pathway. The glyoxylate cycle is found in many bacteria, including *Azotobacter vinelandii* and particularly in organisms that grow well in media in which acetate and other Krebs cycle dicarboxylic acid intermediates are the sole carbon growth source. One primary function of the glyoxylate cycle is to replenish the tricarboxylic and dicarboxylic acid intermediates that are normally provided by the Krebs cycle. A pathway whose primary purpose is to replenish such intermediate compounds is called anaplerotic.

Electron Transport and Oxidative Phosphorylation

The final stage of respiration occurs through a series of oxidation-reduction electron transfer reactions that yield the energy to drive oxidative phosphorylation; this in turn produces ATP. The enzymes involved in electron transport and oxidative phosphorylation reside on the bacterial inner (cytoplasmic) membrane. This membrane is invaginated to form structures called respiratory vesicles, lamellar vesicles, or mesosomes, which function as the bacterial equivalent of the eukaryotic mitochondrial membrane.

Respiratory electron transport chains vary greatly among bacteria, and in some organisms are absent. The respiratory electron transport chain of eukaryotic mitochondria oxidizes NADH + H+, NADPH + H+, and succinate (as well as the coacylated fatty acids such as acetyl~SCoA). The bacterial electron transport chain also oxidizes these compounds, but it can also directly oxidize, via non-pyridine nucleotide-dependent pathways, a larger variety of reduced substrates such as lactate, malate, formate, a-glycerophosphate, H_2, and glutamate. The respiratory electron carriers in bacterial electron transport systems are more varied than in eukaryotes, and the chain is usually branched at the site(s) reacting with molecular O_2. Some electron carriers, such as nonheme iron centers and ubiquinone (coenzyme Q), are common to both the bacterial and mammalian respiratory electron transport chains. In some bacteria, the naphthoquinones or vitamin K may be found with ubiquinone. In still other bacteria, vitamin K serves in the absence of ubiquinone. In mitochondrial respiration, only one cytochrome oxidase component is found (cytochrome *a* + *a3* oxidase). In bacteria there are multiple cytochrome oxidases, including cytochromes *a, d, o,* and occasionally *a* + *a3*.

In bacteria cytochrome oxidases usually occur as combinations of *a1: d: o* and *a* + *a3*: *o*. Bacteria also possess mixed-function oxidases such as cytochromes P-450 and P-420 and cytochromes *c'* and *c'c'*, which also react with carbon monoxide. These diverse types of oxygen-reactive cytochromes undoubtedly have evolutionary significance. Bacteria were present before O_2 was formed; when O_2 became available as a metabolite, bacteria evolved to use it in different ways; this probably accounts for the diversity in bacterial oxygen-reactive hemoproteins.

Cytochrome oxidases in many pathogenic bacteria are studied by the bacterial oxidase reaction, which subdivides Gram-negative organisms into two major groups, oxidase positive and oxidase negative. This oxidase reaction is assayed for by using N,N,N', N'-tetramethyl-*p*-phenylenediamine oxidation (to Wurster's blue) or by using indophenol blue synthesis (with dimethyl-*p*-phenylenediamine and a-naphthol). Oxidase-positive bacteria contain integrated (cytochrome *c* type:oxidase) complexes, the oxidase component most frequently encountered is cytochrome *o*, and occasionally *a* + *a3*. The cytochrome oxidase responsible for the indophenol oxidase reaction complex was isolated from membranes of *Azotobacter vinelandii*, a bacterium with the highest respiratory rate of any known cell. The cytochrome oxidase was found to be an integrated cytochrome *c4:o complex*, which was shown to be present in *Bacillus* species. These *Bacillus* strains are also highly oxidase positive, and most are found in morphologic group II.

Both bacterial and mammalian electron transfer systems can carry out electron transfer (oxidation) reactions with NADH + H^+, NADPH + H^+, and succinate. Energy generated from such membrane oxidations is conserved within the membrane and then transferred in a coupled manner to drive the formation of ATP. The electron transfer sequence is accomplished entirely by membrane-bound enzyme systems. As the electrons are transferred by a specific sequence of electron carriers, ATP is synthesized from ADP + inorganic phosphate (Pi) or orthophosphoric acid (H3PO4).

In respiration, the electron transfer reaction is the primary mode of generating energy; electrons (2*e*-) from a low-redox-potential compound such as NADH + H+ are sequentially transferred to a specific flavoprotein dehydrogenase or oxidoreductase (flavin mononucleotide [FMN] type for NADH or flavin adenine dinucleotide [FAD] type for succinate); this electron pair is then transferred to a nonheme iron center (FeS) and finally to a specific ubiquinone or a naphthoquinone derivative. This transfer of electrons causes a differential chemical redox potential change so that within the membrane enough chemical energy is conserved to be transferred by a coupling mechanism to a high-energy compound (e.g., ADP + Pi Æ ATP). ATP molecules represent the final stable high-energy intermediate compound formed.

A similar series of redox changes also occurs between ubiquinone and cytochrome c, but with a greater differential in the oxidation-reduction potential level, which allows for another ATP synthesis step. The final electron transfer reaction occurs at the cytochrome oxidase level between reduced cyotchrome c and molecular O_2; this reaction is the terminal ATP synthesis step.

MITCHELL OR PROTON EXTRUSION HYPOTHESIS

A highly complex but attractive theory to explain energy conservation in biologic systems is the chemiosmotic coupling of oxidative and photosynthetic phosphorylations, commonly called the Mitchell hypothesis. This theory attempts to explain the conservation of free energy in this process on the basis of an osmotic potential caused by a proton concentration differential (or proton gradient) across a proton-impermeable membrane. Energy is generated by a proton extrusion reaction during membrane-bound electron transport, which in essence serve as a proton pump; energy conservation and coupling follow. This represents an obligatory "intact" membrane phenomenon. The energy thus conserved (again within the confines of the membrane and is coupled to ATP synthesis.

This would occur in all biologic cells, even in the lactic acid bacteria that lack a cytochrome-dependent electron transport chain but still possesses a cytoplasmic membrane. In this hypothesis, the membrane allows for charge separation, thus forming a proton gradient that drives all bioenergization reactions. By such means, electromotive forces can be generated by oxidation-reduction reactions that can be directly coupled to ion translocations, as in the separation of H+ and OH ions in electrochemical systems. Thus, an enzyme or an electron transfer carrier on a membrane that undergoes an oxidation-reduction reaction serves as a specific conductor for OH (or 0_2), and "hydrodehydration" provides electromotive power, as it does in electrochemical cells.

The concept underlying Mitchell's hypothesis is complex, and many modifications have been proposed, but the theory's most attractive feature is that it unifies all bioenergetic conservation principles into a single concept requiring an intact membrane vesicle to function properly. Figure 4-9 shows how the Mitchell hypothesis might be used to explain energy generation, conservation, and transfer by a coupling process. The least satisfying aspect of the chemiosmotic hypothesis is the lack of understanding of how chemical energy is actually conserved within the membrane and how it is transmitted by coupling for ATP synthesis.

Bacterial Photosynthesis

Many prokaryotes (bacteria and cyanobacteria) possess phototrophic modes of metabolism. The types of photosynthesis in the two groups of prokaryotes differ mainly in the type of compound that serves as the hydrogen donor in the reduction of CO_2 to glucose. Phototrophic organisms differ from heterotrophic organisms in that they utilize the glucose synthesized intracellularly for biosynthetic purposes (as in starch synthesis) or for energy production, which usually occurs through cellular respiration.

Unlike phototrophs, heterotrophs require glucose (or some other preformed organic compound) that is directly supplied as a substrate from an exogenous source. Heterotrophs cannot synthesize large concentrations of glucose from CO_2 by specifically using H_2O or (H2S) as a hydrogen source and sunlight as energy. Plant metabolism is a classic example of photolithotrophic metabolism: plants need CO_2 and sunlight; H_2O must be provided as a hydrogen source and usually NO_3 is the nitrogen source for protein synthesis. Organic nitrogen, supplied as fertilizer, is converted to NO_3 in all soils by bacteria via the process of ammonification and nitrification. Although plant cells are phototrophic, they also exhibit a heterotrophic mode of metabolism in that they respire. For example, plants use classic respiration to catabolize glucose that is generated photosynthetically.

Mitochondria as well as the soluble enzymes of the glycolytic pathway are required for glucose dissimilation, and these enzymes are also found in all plant cells. The soluble Calvin cycle enzymes, which are required for glucose synthesis during photosynthesis, are also found in plant cells. It is not possible to feed a plant by pouring a glucose solution on it, but water supplied to a plant will be "photolysed" by chloroplasts in the presence of light; the hydrogen(s) generated from H_2O is used by Photosystems I and II (PSI and PSII) to reduce NADP+ to NADPH + H+. With the ATP generated by PSI and PSII, these reduced pyridine nucleotides, CO_2 is reduced intracellularly to glucose. This metabolic process is carried out in an integrated manner by Photosystems I and II ("Z" scheme) and by the Calvin cycle pathway. A new photosynthetic, and nitrogen fixing bacterium, *Heliobacterium chlorum*, staining Gram positive was isolated, characterized, and found to contain a new type of chlorophyll, i.e., bacteriochlorophyll 'g'. 16S r-RNA sequence analyses showed this organism to be phylogenetically related to members of the family *Bacillaceae*, although all currently known phototrophes are Gram negative.

A few *Heliobacteriium* strains did show the presence of endospores. Another unusual phototrophe is the Gram negative *Halobacterium halobium* (now named *Halobacterium salinarium*), an archaebacterium growing best at 30°C in 4.0-5.0 M (or 25%, w/v) NaCl. This bacterium is a facultative phototrophe having a respiratory mode; it also possesses a purple membrane within which bacteriorhodopsin serves as the active photosynthetic pigment. This purple membranae possesses a light driven proton translocation pump which mediates photosynthetic ATP synthesis via a proton extrusion reaction (see Mitchell Hypothesis).

Autotrophy

Bacteria that grow solely at the expense of inorganic compounds (mineral ions), without using sunlight as an energy source, are called autotrophs, chemotrophs, chemoautotrophs, or chemolithotrophs. Like photosynthetic organisms, all autotrophs use CO_2 as a carbon source for growth; their nitrogen comes from inorganic compounds such as NH_3, NO_3^-, or N_2. Interestingly, the energy source for such organisms is the oxidation of specific inorganic compounds. Which inorganic compound is oxidized depends on the bacteria in question. Many autotrophs will not grow on media that contain organic matter, even agar.

Also found among the autotrophic microorganisms are the sulfur-oxidizing or sulfur-compound-oxidizing bacteria, which seldom exhibit a strictly autotrophic mode of metabolism like the obligate nitrifying bacteria (see discussion of nitrogen cycle below). The representative sulfur compounds oxidized by such bacteria are H_2S, S_2, and S_2O_3. Among the sulfur bacteria are two very interesting organisms; *Thiobacillus ferrooxidans*, which gets its energy for autotrophic growth by oxidizing elemental sulfur or ferrous iron, and *T denitrificans*, which gets its energy by oxidizing S_2O_3 anaerobically, using NO_3^- as the sole terminal electron acceptor. T denitrificans reduces NO_3 to molecular N_2, which is liberated as a gas; this biologic process is called denitrification.

All autotrophic bacteria must assimilate CO_2, which is reduced to glucose from which organic cellular matter is synthesized. The energy for this biosynthetic process is derived from the oxidation of inorganic compounds discussed in the previous paragraph. Note that all autotrophic and phototrophic bacteria possess essentially the same organic cellular constituents found in heterotrophic bacteria; from a nutritional viewpoint, however, the autotrophic mode of metabolism is unique, occurring only in bacteria.

Anerobic Respiration

Some bacteria exhibit a unique mode of respiration called anaerobic respiration. These heterotrophic bacteria that will not grow anaerobically unless a specific chemical component, which serves as a terminal electron acceptor, is added to the medium. Among these electron acceptors are NO_3, SO_4^2, the organic compound fumarate, and CO_2. Bacteria requiring one of these compounds for anaerobic growth are said to be anaerobic respirers.

A large group of anaerobic respirers are the nitrate reducers. The nitrate reducers are predominantly heterotrophic bacteria that possess a complex electron transport system(s) allowing the NO3 ion to serve anaerobically as a terminal acceptor of electrons (NO_3 2e- NO_2; NO_3 5e- N_2; or NO_3 8e- NH_3).The nitrate reductase activity is common in bacteria and is routinely used in the simple nitrate reductase test to identify bacteria.

The methanogens are among the most anaerobic bacteria known, being very sensitive to small concentrations of molecular O_2. They are also archaebacteria, which typically live in unusual and deleterious environments.

All of the above anaerobic respirers obtain chemical energy for growth by using these anaerobic energy-yielding oxidation reactions.

The Nitrogen Cycle

Nowhere can the total metabolic potential of bacteria and their diverse chemical-transforming capabilities be more fully appreciated than in the geochemical cycling of the element nitrogen. All the basic chemical elements (S, O, N, C, and H) required to sustain living organisms have geochemical cycles similar to the nitrogen cycle.

The nitrogen cycle is an ideal demonstration of the ecologic interdependence of bacteria, plants, and animals. Nitrogen is recycled when organisms use one form of nitrogen for growth and excrete another nitrogenous compound as a waste product. This waste product is in turn utilized by another type of organism as a growth or energy substrate.

When the specific breakdown of organic nitrogenous compounds occurs, that is, when proteins are degraded to amino acids (proteolysis) and then to inorganic NH3, by heterotrophic bacteria, the process is called ammonification. This is an essential step in the nitrogen cycle. At death, the organic constituents of the tissues and cells decompose biologically to inorganic constituents by a process called mineralization; these inorganic end products can then serve as nutrients for other life forms. The NH3 liberated in turn serves as a utilizable nitrogen source for many other bacteria. The breakdown of feces and urine also occurs by ammonification.

The other important biologic processes in the nitrogen cycle include nitrification (the conversion of NH_3 to NO_3 by autotrophes in the soil; denitrification (the anaerobic conversion of NO_3 to N_2 gas) carried out by many heterotrophs); and nitrogen fixation (N_2 to NH_3, and cell protein). The latter is a very specialized prokaryotic process called diazotrophy, carried out by both free-living bacteria (such as *Azotobacter, Derxia, Beijeringeia,* and *Azomona* species) and symbionts (such as *Rhizobium* species) in conjunction with legume plants (such as soybeans, peas, clover, and bluebonnets). All plant life relies heavily on NO3- as a nitrogen source, and most animal life relies on plant life for nutrients.

4

Genetics

INTRODUCTION

Genetic Information In Microbes

The genetic material of bacteria and plasmids is DNA. Bacterial viruses (bacteriophages or phages) have DNA or RNA as genetic material. The two essential functions of genetic material are replication and expression. Genetic material must replicate accurately so that progeny inherit all of the specific genetic determinants (the genotype) of the parental organism. Expression of specific genetic material under a particular set of growth conditions determines the observable characteristics (phenotype) of the organism. Bacteria have few structural or developmental features that can be observed easily, but they have a vast array of biochemical capabilities and patterns of susceptibility to antimicrobial agents or bacteriophages. These latter characteristics are often selected as the inherited traits to be analyzed in studies of bacterial genetics.

Nucleic Acid Structure

Nucleic acids are large polymers consisting of repeating nucleotide units. Each nucleotide contains one phosphate group, one pentose or deoxypentose sugar, and one purine or pyrimidine base. In DNA the sugar is D-2-deoxyribose; in RNA the sugar is D-ribose. In DNA the purine bases are adenine (A) and guanine (G), and the pyrimidine bases are thymine (T) and cytosine (C). In RNA, uracil (U) replaces thymine. Chemically modified purine and pyrimidine bases are found in some bacteria and bacteriophages. The repeating structure of polynucleotides involves alternating sugar and phosphate residues, with phosphodiester bonds linking the 3'-hydroxyl group of one nucleotide sugar to the 5'-hydroxyl group of the adjacent nucleotide sugar. These asymmetric phosphodiester linkages define the polarity of the polynucleotide chain.

A purine or pyrimidine base is linked at the 1'-carbon atom of each sugar residue and projects from the repeating sugar-phosphate backbone. Double-stranded DNA is helical, and the two strands in the helix are antiparallel. The double helix is stabilized by hydrogen bonds between purine and pyrimidine bases on the opposite strands. At each position, A on one strand pairs by two hydrogen bonds with T on the opposite strand, or G pairs by three hydrogen bonds with C. The two strands of double-helical DNA are, therefore, complementary. Because of complementarity, double-stranded DNA contains equimolar amounts of purines (A + G) and pyrimidines (T + C), with A equal to T and G equal to C, but the mole fraction of G + C in DNA varies widely among different bacteria. Information in nucleic acids is encoded by the ordered sequence of nucleotides along the polynucleotide chain, and in double-stranded DNA the sequence of each strand determines what the sequence of the complementary strand must be. The extent of sequence homology between DNAs from different microorganisms is the most stringent criterion for determining how closely they are related.

DNA Replication

During replication of the bacterial genome, each strand in double-helical DNA serves as a template for synthesis of a new complementary strand. Each daughter double-stranded DNA molecule thus contains one old polynucleotide strand and one newly synthesized strand. This type of DNA replication is called semiconservative. Replication of chromosomal DNA in bacteria starts at a specific chromosomal site called the origin and proceeds bidirectionally until the process is completed. When bacteria divide by binary fission after completing DNA replication, the replicated chromosomes are partitioned into each of the daughter cells. The origin regions specifically and transiently associate with the cell membrane after DNA replication has been intitiated, leading to a model whereby membrane attachment directs separation of daughter chromosomes (the replicon model). These characteristics of DNA replication during bacterial growth fulfill the requirements of the genetic material to be reproduced accurately and to be inherited by each daughter cell at the time of cell division.

Gene Expression

Genetic information encoded in DNA is expressed by synthesis of specific RNAs and proteins, and information flows from DNA to RNA to protein. The DNA-directed synthesis of RNA is called transcription. Because the strands of double-helical DNA are antiparallel and complementary, only one of the two DNA strands can serve as template for synthesis of a specific mRNA molecule. Messenger RNAs (mRNAs) transmit information from DNA, and each mRNA in bacteria functions as the template for synthesis of one or more specific proteins. The process by which the nucleotide sequence of an mRNA molecule determines the primary amino acid sequence of a protein is called translation. Ribosomes, complexes of ribosomal RNAs (rRNAs) and several ribosomal proteins, translate each mRNA into the corresponding polypeptide sequence with the aid of transfer RNAs (tRNAs), amino-acyl tRNA synthesases, initiation factors and elongation factors. All of these components of the apparatus for protein synthesis function in the production of many different proteins. A gene is a DNA sequence that encodes a protein, rRNA, or tRNA molecule (gene product).

The genetic code determines how the nucleotides in mRNA specify the amino acids in a polypeptide. Because there are only 4 different nucleotides in mRNA (containing U, A, C and G), single nucleotides do not contain enough information to specify uniquely all 20 of the amino acids. In dinucleotides 16 (4 x 4) arrangements of the four nucleotides are possible, and in trinucleotides 64 (4 x 4 x 4) arrangements are possible. Thus, a minimum of three nucleotides is required to provide at least one unique sequence corresponding to each of the 20 amino acids. The "universal" genetic code employed by most organisms (Table 1) is a triplet code in which 61 of the 64 possible trinucleotides (codons) encode specific amino acids, and any of the three remaining codons (UAG, UAA or UGA) results in termination of translation. The chain-terminating codons are also called nonsense codons because they do not specify any amino acids.

The genetic code is described as degenerate, because several codons may be used for a single amino acid, and as nonoverlapping, because adjacent codons do not share any common nucleotides. Exceptions to the "universal" code include the use of UGA as a tryptophan codon in some species of Mycoplasma and in mitochondrial DNA, and a few additional codon differences in mitochondrial DNAs from yeasts, Drosophila, and mammals. Translation of mRNA is usually initiated at an AUG codon for methionine, and adjacent codons are translated sequentially as the mRNA is read in the 5' to 3' direction. The corresponding polypeptide chain is assembled beginning at its amino terminus and proceeding toward its carboxy terminus. The sequence of amino acids in the polypeptide is, therefore, colinear with the sequence of nucleotides in the mRNA and the corresponding gene. Specific enzymatic reactions involved in DNA, RNA, and protein synthesis are beyond the scope of this chapter.

Expression of genetic determinants in bacteria involves the unidirectional flow of information from DNA to RNA to protein. In bacteriophages, either DNA or RNA can serve as genetic material. During infection of bacteria by RNA bacteriophages, RNA molecules serve as templates for RNA replication and as mRNAs. Studies with the retrovirus group of animal viruses reveal that DNA molecules can be synthesized from RNA templates by enzymes designated as RNA-dependent DNA polymerases (reverse transcriptases). This reversal of the usual direction for flow of genetic information, from RNA to DNA instead of from DNA to RNA, is an important mechanism for enabling information from retroviruses to be encoded in DNA and to become incorporated into the genomes of animal cells.

Genome Organization

DNA molecules that replicate as discrete genetic units in bacteria are called replicons. In some Escherichia coli strains, the chromosome is the only replicon present in the cell. Other bacterial strains have additional replicons, such as plasmids and bacteriophages.

Chromosomal DNA

Bacterial genomes vary in size from about 0.4 x 109 to 8.6 x 109 daltons (Da), some of the smallest being obligate parasites (Mycoplasma) and the largest belonging to bacteria capable of complex differentiation such as Myxococcus. The amount of DNA in the genome determines the maximum amount of information that it can encode. Most bacteria have a haploid genome, a single chromosome consisting of a circular, double stranded DNA molecule. However linear chromosomes have been found in Gram-positive Borrelia and Streptomyces spp., and one linear and one circular chromosome is present in the Gram-negative bacterium Agrobacterium tumefaciens. The single chromosome of the common intestinal bacterium E coli is 3 x 109 Da (4,500 kilobase pairs [kbp]) in size, accounting for about 2 to 3 percent of the dry weight of the cell. The E coli genome is only about 0.1% as large as the human genome, but it is sufficient to code for several thousand polypeptides of average size (40 kDa or 360 amino acids).

The chromosome of E coli has a contour length of approximately 1.35 mm, several hundred times longer than the bacterial cell, but the DNA is supercoiled and tightly packaged in the bacterial nucleoid. The time required for replication of the entire chromosome is about 40 minutes, which is approximately twice the shortest division time for this bacterium. DNA replication must be initiated as often as the cells divide, so in rapidly growing bacteria a new round of chromosomal replication begins before an earlier round is completed. At rapid growth rates there may be four chromosomes replicating to form eight at the time of cell division, which is coupled with completion of a round of chromosomal replication. Thus, the chromosome in rapidly growing bacteria is replicating at more than one point. The replication of chromosomal DNA in bacteria is complex and involves many different proteins.

Plasmids

Plasmids are replicons that are maintained as discrete, extrachromosomal genetic elements in bacteria. They are usually much smaller than the bacterial chromosome, varying from less than 5 to more than several hundred kbp, though plasmids as large as 2 Mbp occur in some bacteria. Plasmids usually encode traits that are not essential for bacterial viability, and replicate independently of the chromosome. Most plasmids are supercoiled, circular, double-stranded DNA molecules, but linear plasmids have also been demonstrated in Borrelia and Streptomyces. Closely related or identical plasmids demonstrate incompatibility; they cannot be stably maintained in the same bacterial host. Classification of plasmids is based on incompatibility or on use of specific DNA probes in hybridization tests to identify nucleotide sequences that are characteristic of specific plasmid replicons. Some hybrid plasmids contain more than one replicon.

Conjugative plasmids code for functions that promote transfer of the plasmid from the donor bacterium to other recipient bacteria, but nonconjugative plasmids do not. Conjugative plasmids that also promote transfer of the bacterial chromosome from the donor bacterium to other recipient bacteria are called fertility plasmids, and are discussed below. The average number of molecules of a given plasmid per bacterial chromosome is called its copy number. Large plasmids (>40 kilobase pairs) are often conjugative, have small copy numbers (1 to several per chromosome), code for all functions required for their replication, and partition themselves among daughter cells during cell division in a manner similar to the bacterial chromosome.

Plasmids smaller than 7.5 kilobase pairs usually are nonconjugative, have high copy numbers (typically 10-20 per chromosome), rely on their bacterial host to provide some functions required for replication, and are distributed randomly between daughter cells at division.

Many plasmids control medically important properties of pathogenic bacteria, including resistance to one or several antibiotics, production of toxins, and synthesis of cell surface structures required for adherence or colonization. Plasmids that determine resistance to antibiotics are often called R plasmids (or R factors). Representative toxins encoded by plasmids include heat-labile and heat-stable enterotoxins of E coli, exfoliative toxin of Staphylococcus aureus, and tetanus toxin of Clostridium tetani. Some plasmids are cryptic and have no recognizable effects on the bacterial cells that harbor them. Comparing plasmid profiles is a useful method for assessing possible relatedness of individual clinical isolates of a particular bacterial species for epidemiological studies. The role of plasmids in the evolution of resistance to antibiotics is discussed below.

Bacteriophages

Bacteriophages (bacterial viruses, phages) are infectious agents that replicate as obligate intracellular parasites in bacteria. Extracellular phage particles are metabolically inert and consist principally of proteins plus nucleic acid (DNA or RNA, but not both). The proteins of the phage particle form a protective shell (capsid) surrounding the tightly packaged nucleic acid genome. Phage genomes vary in size from approximately 2 to 200 kilobases per strand of nucleic acid and consist of double-stranded DNA, single-stranded DNA, or RNA. Phage genomes, like plasmids, encode functions required for replication in bacteria, but unlike plasmids they also encode capsid proteins and nonstructural proteins required for phage assembly. Several morphologically distinct types of phage have been described, including polyhedral, filamentous, and complex. Complex phages have polyhedral heads to which tails and sometimes other appendages (tail plates, tail fibers, etc.) are attached.

The capsids remain at the cell surface, and the DNA or RNA genomes enter the target cells (penetration). Because infectivity of genomic DNA or RNA is much less than that of mature virus, there is a time immediately after infection called the eclipse period during which intracellular infectious phage cannot be detected. The infecting phage RNA or DNA is replicated to produce many new copies of the phage genome, and phage-specific proteins are produced. For most phages assembly of progeny occurs in the cytoplasm, and release of the progeny occurs by cell lysis. In contrast, filamentous phages are formed at the cell envelope and released without killing the host cells. The eclipse period ends when intracellular infectious progeny appear. The latent period is the interval from infection until extracellular progeny appear, and the rise period is the interval from the end of the latent period until all phage are extracellular. The average number of phage particles produced by each infected cell, called the burst size, is characteristic for each virus and often ranges between 50 and several hundred.

Phages are classified into two major groups: virulent and temperate. Growth of virulent phages in susceptible bacteria destroys the host cells. Infection of susceptible bacteria by temperate phages can have either of two outcomes: lytic growth or lysogeny. Lytic growth of temperate and virulent bacteriophages is similar, leading to production of phage progeny and death of the host bacteria. Lysogeny is a specific type of latent viral infection in which the phage genome replicates as a prophage in the bacterial cell. In most lysogenic bacteria the genes required for lytic phage development are not expressed, and production of infectious phage does not occur. Furthermore, the lysogenic cells are immune to superinfection by the virus which they harbor as a prophage. The physical state of the prophage is not identical for all temperate viruses. For example, the prophage of bacteriophage l in E coli is integrated into the bacterial chromosome at a specific site and replicates as part of the bacterial chromosome, whereas the prophage of bacteriophage P1 in E coli replicates as an extrachromosomal plasmid.

Lytic phage growth occurs spontaneously in a small fraction of lysogenic cells, and a few extracellular phages are present in cultures of lysogenic bacteria. For some lysogenic bacteria, synchronous induction of lytic phage development occurs in the entire population of lysogenic bacteria when they are treated with agents that damage DNA, such as ultraviolet light or mitomycin C. The loss of prophage from a lysogenic bacterium, converting it to the nonlysogenic state and restoring susceptibility to infection by the phage that was originally present as prophage, is called curing.

Some temperate phages contain genes for bacterial characteristics that are unrelated to lytic phage development or the lysogenic state, and expression of such genes is called phage conversion (or lysogenic conversion). Examples of phage conversion that are important for microbial virulence include production of diphtheria toxin by Corynebacterium diphtheriae, erythrogenic toxin by Streptococcus pyogenes (group A b-hemolytic streptococci), botulinum toxin by Clostridium botulinum, and Shiga-like toxins by E coli. In each of these examples the gene which encodes the bacterial toxin is present in a temperate phage genome. The specificity of O antigens in Salmonella can also be controlled by phage conversion. Phage typing is the testing of strains of a particular bacterial species for susceptibility to specific bacteriophages. The patterns of susceptibility to the set of typing phages provide information about the possible relatedness of individual clinical isolates. Such information is particularly useful for epidemiological investigations.

Mutation and Selection

Mutations are heritable changes in the genome. Spontaneous mutations in individual bacteria are rare. Some mutations cause changes in phenotypic characteristics; the occurrence of such mutations can be inferred from the effects they produce. In microbial genetics specific reference organisms are designated as wild-type strains, and descendants that have mutations in their genomes are called mutants. Thus, mutants are characterized by the inherited differences between them and their ancestral wild-type strains. Variant forms of a specific genetic determinant are called alleles. Genotypic symbols are lower case, italicized abbreviations that specify individual genes, with a (+) superscript indicating the wild type allele. Phenotypic symbols are capitalized and not italicized, to distinguish them from genotypic symbols. For example, the genotypic symbol for the ability to produce b-galactosidase, required to ferment lactose, is lacZ+, and mutants that cannot produce b-galactosidase are lacZ. The lactose-fermenting phenotype is designated Lac+, and inability to ferment lactose is Lac-.

DETECTION OF MUTANT PHENOTYPES

Selective and differential media are helpful for isolating bacterial mutants. Some selective media permit particular mutants to grow, but do not allow the wild-type strains to grow. Rare mutants can be isolated by using such selective media. Differential media permit wild-type and mutant bacteria to grow and form colonies that differ in appearance. Detection of rare mutants on differential media is limited by the total number of colonies that can be observed. Consider a wild-type strain of E coli that is susceptible to the antibiotic streptomycin (phenotype Strs) and can utilize lactose as the sole source of carbon (phenotype Lac+). Spontaneously occurring Strr mutants are rare and are usually found at frequencies of less than one per 109 bacteria in cultures of wild-type E coli.

Nevertheless, Strr mutants can be isolated easily by using selective media containing streptomycin, because the wild-type Strs bacteria are killed. Isolation of lactose-negative (phenotype Lac-) mutants of E coli poses a different problem. On minimal media with lactose as the sole source of carbon, Lac+ wild-type strains will grow, but Lac- mutants cannot grow. On differential media such as MacConkey-lactose agar or eosin-methylene blue-lactose agar, Lac+ wild-type and Lac- mutant strains of E coli can be distinguished by their color, but spontaneous Lac- mutants are too rare to be isolated easily. Selective media for Lac- mutants of E coli can be made by incorporating chemical analogs of lactose that are converted into toxic metabolites by Lac+ bacteria but not by Lac- mutants. The Lac- mutants can then grow on such media, but the Lac+ wild-type bacteria are killed.

Mutations that inactivate essential genes in haploid organisms are usually lethal, but such potentially lethal mutations can often be studied if their expression is controlled by manipulation of experimental conditions. For example, a mutation that increases the thermolability of an essential gene product may prevent bacterial growth at 42°C, although the mutant bacterium can still grow at 25°C. Conversely, cold-sensitive mutants express the mutant phenotype at low temperature, but not at high temperature. Temperature-sensitive and cold-sensitive mutations are examples of conditional mutations, as are suppressible mutations described later in this chapter. A conditional lethal phenotype indicates that the mutant gene is essential for viability.

Spontaneous and Induced Mutations

The mutation rate in bacteria is determined by the accuracy of DNA replication, the occurrence of damage to DNA, and the effectiveness of mechanisms for repair of damaged DNA.

For a particular bacterial strain under defined growth conditions, the mutation rate for any specific gene is constant and is expressed as the probability of mutation per cell division. In a population of bacteria grown from a small inoculum, the proportion of mutants usually increases progressively as the size of the bacterial population increases.

Mutations in bacteria can occur spontaneously and independently of the experimental methods used to detect them. This principle was first demonstrated by the fluctuation test. The numbers of phage-resistant mutants of E coli in replicate cultures grown from small inocula were measured and compared with those in multiple samples taken from a single culture. If mutations to phage resistance occurred only after exposure to phage, the variability in numbers of mutants between cultures should be similar under both sets of conditions. In contrast, if phage-resistant mutants occurred spontaneously before exposure of the bacteria to phage, the numbers of mutants should be more variable in the independently grown cultures, because differences in the size of the bacterial population when the first mutant appeared would contribute to the observed variability. The data indicated that the mutations to phage resistance in E coli occurred spontaneously with constant probability per cell division.

Replica plating confirmed that mutations in bacteria can occur spontaneously, without exposure of bacteria to selective agents. For replica plating, a flat, sterile, velveteen surface is used to pick up an inoculum from the surface of an agar master plate and transfer samples to other agar plates. In this manner, samples of the bacterial population from the master plate are transferred to the replica plates without distorting their spatial arrangement. If the replica plates contain selective medium and the master plates do not, the positions of selected mutant colonies on the replica plates can be noted, and bacteria that were not exposed to the selective conditions can be isolated from the same positions on the master plate. Mutants of E coli resistant to bacteriophage T1 or to streptomycin have been isolated in this way, without exposing the wild-type bacteria to the bacteriophage or the antibiotic.

Both environmental and genetic factors affect mutation rates. Exposure of bacteria to mutagenic agents causes mutation rates to increase, sometimes by several orders of magnitude. Many chemical and physical agents, including X-rays and ultraviolet light, have mutagenic activity. Chemicals that are carcinogenic for animals are often mutagenic for bacteria, or can be converted by animal tissues to metabolites that are mutagenic for bacteria. Standardized tests for mutagenicity in bacteria are used as screening procedures to identify environmental agents that may be carcinogenic in humans. Mutator genes in bacteria cause an increase in spontaneous mutation rates for a wide variety of other genes. Expression of these genes, induced by DNA damage (see SOS response later), enables the repair of DNA lesions that would otherwise be lethal, but by an error-prone mechanism that increases the rate of mutation. The overall mutation ratethe probability that a mutation will occur somewhere in the bacterial genome per cell divisionis relatively constant for a variety of organisms with genomes of different sizes and appears to be a significant factor in determining the fitness of a bacterial strain for survival in nature. Most mutations are deleterious, and the risk of adverse mutations for individual bacteria must be balanced against the positive value of mutability as a mechanism for adaptation of bacterial populations to changing environmental conditions.

Molecular Basis of Mutations

Mutations are classified on the basis of structural changes that occur in DNA. Some mutations are localized within short segments of DNA (for example, nucleotide substitutions, microdeletions, and microinsertions). Other mutations involve large regions of DNA and include deletions, insertions, or rearrangements of segments of DNA.

When a nucleotide substitution occurs in a region of DNA that codes for a polypeptide, one of the three nucleotides within a single codon of a corresponding mRNA molecule will be changed. Silent mutations cause no change in polypeptide structure or function, because one codon in mRNA is changed to another for the same amino acid. Other substitutions cause one amino acid to be replaced by another at the specific position within the polypeptide corresponding to the altered codon. Mutations that result in replacement of one amino acid for another within a polypeptide chain are called missense mutations. The effects of amino acid replacements on the function of a polypeptide gene product vary and depend on the location and the identity of the amino acid replacement. Mutant polypeptides containing amino acid replacements usually share antigenic determinants with the wild-type polypeptide and often have some residual biologic activity. Mutations that result in replacement of an amino acid codon with a termination codon are called nonsense mutations. This results in production of an amino-terminal fragment of the normal polypeptide when the mutant mRNA is translated. Nonsense mutations often result in complete loss of activity of the gene product.

Because of the triplet nature of the genetic code, the consequences of mutations caused by insertions or deletions of small numbers of nucleotides (microinsertions, microdeletions) depend on both the number and sequence of nucleotides involved. Deletion or addition of multiples of three nucleotide pairs does not affect the reading frame, but causes deletion or addition of appropriate numbers of amino acids at one site within the polypeptide. If a new chain-terminating codon is introduced, premature chain termination occurs within the polypeptide. In contrast, addition or deletion of other numbers of nucleotide pairs alters the reading frame for the entire segment of mRNA from the mutation to the distal end of the gene. Therefore, frameshift mutations are likely to cause drastic changes in the structure and activity of polypeptide gene products, and they are often classified as nonsense mutations.

Complementation Tests

To determine if mutations are located in the same gene or different genes, complementation tests are performed with partially diploid bacterial strains. Two copies of the region of the bacterial chromosome harboring a mutation are present in the same bacterium, with each copy containing a different mutation (mutations are in the trans arrangement). A wild-type phenotype indicates that the mutations are in different genes. This phenomenon is called complementation. If a mutant phenotype is observed, a control experiment should be performed with the mutations in the cis arrangement to exclude the possibility that the wild-type alleles cannot be expressed normally in a partially diploid bacterial strain. Complementation tests were originally called "cis-trans" tests, and the term cistron is sometimes used as a synonym for gene. Complementation tests can be performed and interpreted even if the specific biochemical functions of the gene products are unknown.

As an example, consider using a complementation test to characterize two independently derived Lac- mutants of E coli. The biochemical pathway for utilization of lactose requires ß-galactoside permease (genotypic symbol lacY) to transport lactose into the bacterial cell and b-galactosidase (genotypic symbol lacZ) to convert lactose into D-glucose and D-galactose. Mutants that lack b-galactoside permease or b-galactosidase cannot utilize lactose for growth. If the mutations in both Lac-mutants inactivated the same protein (e.g., b-galactoside) then a partial diploid strain containing the lacZ genes from both mutants in the trans arrangement would be unable to utilize lactose. In contrast, if the genotypes of the two mutants were lacZ+ lacY and lacZ lacY+, the partially diploid bacterium would produce active b-galactosidase from the lacZ+ determinant and active b-galactoside permease from the lacY+ determinant. Complementation would occur, and the partially diploid strain would utilize lactose.

Reversion and Suppression

Mutations that convert the phenotype from wild-type to mutant are called forward mutations, and mutations that change the phenotype from mutant back to wild-type are called reverse mutations (reversions). Bacterial strains that contain reverse mutations are called revertants. Analysis of mutations that cause phenotypic reversion yields useful information. Reverse mutations that restore the exact nucleotide sequence of the wild-type DNA are true reversions. True revertants are identical to wild-type strains both genotypically and phenotypically. Reverse mutations that do not restore the exact nucleotide sequence of the wild-type DNA are called suppressor mutations (suppressors). Some revertants that harbor suppressor mutations are phenotypically indistinguishable from wild-type strains. Other revertants, called pseudorevertants, can be distinguished phenotypically from wild-type strains, for example, by subtle differences in the characteristics of an enzymatic activity that has been regained (such as specific activity, substrate specificity, kinetic constants, or susceptibility to thermal or chemical inactivation). Recognition of pseudorevertant phenotypes suggests the presence of suppressor mutations.

Suppressor mutations can be intragenic or extragenic. Intragenic suppressors are located in the same gene as the forward mutations that they suppress. The possible locations and nature of intragenic suppressors are determined by the original forward mutation and by the relationships between the primary structure of the gene product an. its biologic activity. Extragenic suppressors are located in different genes from mutations whose effects they suppress. The ability of extragenic suppressors to suppress a variety of independent mutations can be tested. Some extragenic suppressors are specific for particular genes, some are specific for particular codons, and some have other specificity patterns. Extragenic suppressors that reverse the phenotypic effects of chain-terminating codons have been well characterized and found to alter the structure of specific tRNAs. A particular suppressor tRNA can permit a specific chain-terminating codon to be translated, resulting in incorporation of a specific amino acid into the nascent polypeptide at the position corresponding to the chain-terminating codon. In a bacterium that has a chain-terminating mutation and an appropriate extragenic suppressor, translation of the mRNA containing the mutant codon can therefore result in formation of a full-length polypeptide. The biologic activity of the full-length polypeptide formed as a consequence of suppression depends both on the amount of protein made and on the functional consequences of the specific amino acid replacement determined by the suppressor tRNA.

EXCHANGE OF GENETIC INFORMATION

The biologic significance of sexuality in microorganisms is to increase the probability that rare, independent mutations will occur together in a single microbe and be subjected to natural selection. Genetic interactions between microbes enable their genomes to evolve much more rapidly than by mutation alone. Representative phenomena of medical importance that involve exchanges of genetic information or genomic rearrangements include the rapid emergence and dissemination of antibiotic resistance plasmids, flagellar phase variation in Salmonella, and antigenic variation of surface antigens in Neisseria and Borrelia.

Sexual processes in bacteria involve transfer of genetic information from a donor to a recipient and result either in substitution of donor alleles for recipient alleles or addition of donor genetic elements to the recipient genome. Transformation, transduction, and conjugation are sexual processes that use different mechanisms to introduce donor DNA into recipient bacteria. Because donor DNA cannot persist in the recipient bacterium unless it is part of a replicon, recombination between donor and recipient genomes is often required to produce stable, hybrid progeny. Recombination is most likely to occur when the donor and recipient bacteria are from the same or closely related species.

For a recombinant to be detected, its phenotype must be different from both parental phenotypes. Growth or cell division may be required before the recombinant phenotype is expressed. Delay in expression of a recombinant phenotype until a haploid recombinant genome has segregated is called segregation lag, and delay until synthesis of products encoded by donor genes has occurred is called phenotypic lag. Testing for linkage (nonrandom reassortment of parental alleles in recombinant progeny) is possible when the parental bacteria have different alleles for several genes. The donor allele of an unselected gene is more likely to be present in recombinants if it is linked to the selected donor gene than if it is not linked to the selected donor gene. Quantitative analysis of linkage permits construction of genetic maps. The genome of E coli is circular, as determined both by genetic linkage and direct biochemical analysis of chromosomal DNA, and the genetic map is colinear with the physical map of the chromosomal DNA. Genetic and physical mapping are also used to analyze extrachromosomal replicons such as bacteriophages and plasmids.

Many bacteria have restriction-modification (RM) systems, consisting of modifying enzymes that methylate adenine or cytosine residues at specific sequences in their own DNA and corresponding restriction endonucleases that cleave foreign DNA which does not carry the specific modification at the same target sequences. Some restriction enzymes will only cleave DNA that has been methylated at specific sequences. These restriction systems, which may have evolved to protect bacteria against invasion by phages or plasmids, are an important barrier to genetic exchanges between different bacterial strains or species. Recent evidence suggests that plasmid-borne RM systems may be a way for the plasmid to ensure its carriage in a host strain, since cells that lose the plasmid (and the corresponding protective methylase gene) are killed by the action of the more stable restriction enzyme, which attacks the newly replicated but unmodified chromosomal DNA.

Transformation

In transformation, pieces of DNA released from donor bacteria are taken up directly from the extracellular environment by recipient bacteria. Recombination occurs between single molecules of transforming DNA and the chromosomes of recipient bacteria. To be active in transformation, DNA molecules must be at least 500 nucleotides in length, and transforming activity is destroyed rapidly by treating DNA with deoxyribonuclease. Molecules of transforming DNA correspond to very small fragments of the bacterial chromosome. Cotransformation of genes is unlikely, therefore, unless they are so closely linked that they can be encoded on a single DNA fragment.

Transformation was discovered in Streptococcus pneumoniae and occurs in other bacterial genera including Haemophilus, Neisseria, Bacillus, and Staphylococcus. The ability of bacteria to take up extracellular DNA and to become transformed, called competence, varies with the physiologic state of the bacteria. Many bacteria that are not usually competent can be made to take up DNA by laboratory manipulations, such as calcium shock or exposure to a high-voltage electrical pulse (electroporation). In some bacteria (including Haemophilus and Neisseria) DNA uptake depends on the presence of specific oligonucleotide sequences in the transforming DNA, but in others (including Streptococcus pneumoniae) DNA uptake is not sequence-specific. Competent bacteria may also take up intact bacteriophage DNA (transfection) or plasmid DNA, which can then replicate as extrachromosomal genetic elements in the recipient bacteria. In contrast, a piece of chromosomal DNA from a donor bacterium usually cannot replicate in the recipient bacterium unless it becomes part of a replicon by recombination. Historically, characterization of "transforming principle" from S pneumoniae provided the first direct evidence DNA is genetic material.

Transduction

In transduction, bacteriophages function as vectors to introduce DNA from donor bacteria into recipient bacteria by infection. For some phages, called generalized transducing phages, a small fraction of the virions produced during lytic growth are aberrant and contain a random fragment of the bacterial genome instead of phage DNA. Each individual transducing phage carries a different set of closely linked genes, representing a small segment of the bacterial genome. Transduction mediated by populations of such phages is called generalized transduction, because each part of the bacterial genome has approximately the same probability of being transferred from donor to recipient bacteria.

When a generalized transducing phage infects a recipient cell, expression of the transferred donor genes occurs. Abortive transduction refers to the transient expression of one or more donor genes without formation of recombinant progeny, whereas complete transduction is characterized by production of stable recombinants that inherit donor genes and retain the ability to express them. In abortive transduction the donor DNA fragment does not replicate, and among the progeny of the original transductant only one bacterium contains the donor DNA fragment. In all other progeny the donor gene products become progressively diluted after each generation of bacterial growth until the donor phenotype can no longer be expressed. On selective medium upon which only bacteria with the donor phenotype can grow, abortive transductants produce minute colonies that can be distinguished easily from colonies of stable transductants. The frequency of abortive transduction is typically one to two orders of magnitude greater than the frequency of generalized transduction, indicating that most cells infected by generalized transducing phages do not produce recombinant progeny.

Specialized transduction differs from generalized transduction in several ways. It is mediated only by specific temperate phages, and only a few specific donor genes can be transferred to recipient bacteria. Specialized transducing phages are formed only when lysogenic donor bacteria enter the lytic cycle and release phage progeny. The specialized transducing phages are rare recombinants which lack part of the normal phage genome and contain part of the bacterial chromosome located adjacent to the prophage attachment site. Many specialized transducing phages are defective and cannot complete the lytic cycle of phage growth in infected cells unless helper phages are present to provide missing phage functions. Specialized transduction results from lysogenization of the recipient bacterium by the specialized transducing phage and expression of the donor genes. Phage conversion and specialized transduction have many similarities, but the origin of the converting genes in temperate converting phages is unknown.

Conjugation

In conjugation, direct contact between the donor and recipient bacteria leads to establishment of a cytoplasmic bridge between them and transfer of part or all of the donor genome to the recipient. Donor ability is determined by specific conjugative plasmids called fertility plasmids or sex plasmids.

The F plasmid (also called F factor) of E coli is the prototype for fertility plasmids in Gram-negative bacteria. Strains of E coli with an extrachromosomal F plasmid are called F+ and function as donors, whereas strains that lack the F plasmid are F- and behave as recipients. The conjugative functions of the F plasmid are specified by a cluster of at least 25 transfer (tra) genes which determine expression of F pili, synthesis and transfer of DNA during mating, interference with the ability of F+ bacteria to serve as recipients, and other functions. Each F+ bacterium has 1 to 3 F pili that bind to a specific outer membrane protein (the ompA gene product) on recipient bacteria to initiate mating. An intercellular cytoplasmic bridge is formed, and one strand of the F plasmid DNA is transferred from donor to recipient, beginning at a unique origin and progressing in the 5' to 3' direction. The transferred strand is converted to circular double-stranded F plasmid DNA in the recipient bacterium, and a new strand is synthesized in the donor to replace the transferred strand. Both of the exconjugant bacteria are F+, and the F plasmid can therefore spread by infection among genetically compatible populations of bacteria. In addition to the role of the F pili in conjugation, they also function as receptors for donor-specific (male-specific) phages.

The F plasmid in E coli can exist as an extrachromosomal genetic element or be integrated into the bacterial chromosome. Because the F plasmid and the bacterial chromosome are both circular DNA molecules, reciprocal recombination between them produces a larger DNA circle consisting of F plasmid DNA inserted linearly into the chromosome. E coli contains multiple copies of several different genetic elements called insertion sequences (see section on transposons for more detail), at various locations in its chromosome and in the F plasmid. Homologous recombination between insertion sequences in the chromosome and the F plasmid leads to preferential integration of the F plasmid at chromosomal sites where insertion sequences are located. The chromosomal sites where insertion sequences are found vary, however, among strains of E coli.

An E coli strain with an integrated F plasmid retains its ability to function as a donor in conjugal matings. Because donor strains with integrated F factors can transfer chromosomal genes to recipients with high efficiency, they are called Hfr (High frequency recombination) strains. Transfer of single-stranded DNA from an Hfr donor to a recipient begins from the origin within the F plasmid and proceeds as described above, except that the transferred DNA is the hybrid replicon consisting of F plasmid integrated into the bacterial chromosome. Transfer of this entire replicon, including the bacterial chromosome, requires approximately 100 minutes. The identity of the first chromosomal gene to be transferred and the polarity of chromosomal transfer are determined by the site of integration of the F plasmid and its orientation with respect to the bacterial chromosome. Because the mating bacteria usually separate spontaneously before the entire chromosome is transferred, conjugation typically transfers only a fragment of the donor chromosome into the recipient. The probability that a donor gene will enter the recipient bacterium during conjugation decreases, therefore, as its distance from the F origin (and therefore the time of its transfer) increases.

Mating cells can also be broken apart experimentally by subjecting them to strong shearing forces in a mechanical blender; this is called interrupted mating. Formation of recombinant progeny requires recombination between the transferred donor DNA and the genome of the recipient bacterium. Analysis of progeny from matings that are interrupted after different intervals demonstrates which chromosomal genes are transferred first by particular donor strains, the sequential times of entry for genes that are transferred subsequently, and the progressively lower probability that genes transferred later will appear in recombinant progeny. The circularity of the genetic map of E coli was originally deduced from the overlapping, circularly permuted groups of linked genes that were transferred early by individual donor strains in which the F factor was integrated at different chromosomal locations.

In matings between F+ and F- bacteria, only the F plasmid is transferred with high efficiency to recipients. Chromosomal genes are transferred with very low efficiency, and it is the spontaneous Hfr mutants in F+ populations that mediate transfer of donor chromosomal genes. In matings between Hfr and F- strains, the segment of the F plasmid containing the tra region is transferred last, after the entire bacterial chromosome has been transferred. Most recombinants from matings between Hfr and F- cells fail to inherit the entire set of F plasmid genes and are phenotypically F-. In matings between F+ and F- strains, the F plasmid spreads rapidly throughout the bacterial population, and most recombinants are F+.

Integrated F plasmids in Hfr strains can sometimes be excised from the bacterial chromosome. If excision precisely reverses the integration process, F+ cells are produced. On rare occasions, however, excision occurs by recombinations involving insertion sequences or other genes on the bacterial chromosome that are located at some distance from the original integration site. In such cases segments of the bacterial chromosome can become incorporated into hybrid F plasmids that are called F' plasmids. By similar processes, segments of the bacterial chromosome can sometimes become incorporated into R plasmids to produce hybrid R' plasmids. Conjugative R' plasmids can function as fertility plasmids because they can integrate into the bacterial chromosome by homologous recombination and mediate transfer of chromosomal genes during matings with recipient bacteria. F' plasmids, R' plasmids, specialized transducing phages, and recombinant plasmids or phages constructed by gene cloning (described below) are hybrid replicons that can include segments of the bacterial chromosome. Therefore, any of these genetic elements can be used to construct the partially diploid bacterial strains that are required for complementation tests and other purposes.

Conjugation also occurs in Gram-positive bacteria. Gram-positive donor bacteria produce adhesins that cause them to aggregate with recipient cells, but sex pili are not involved. In some Streptococcus species, recipient bacteria produce extracellular sex pheromones that cause the donor phenotype to be expressed by bacteria that harbor an appropriate conjugative plasmid, and the conjugative plasmid prevents the donor cells from producing the corresponding pheromone.

Recombination

Recombination involves breakage and joining of parental DNA molecules to form hybrid, recombinant molecules. Several distinct kinds of recombination have been identified that depend on different features of the participating genomes and require the activities of different gene products. Specific enzymes that act on DNA (for example, exonucleases, endonucleases, polymerases, ligases) participate in recombination. Detailed discussion of the biochemical events in recombination is beyond the scope of this chapter.

Generalized recombination involves donor and recipient DNA molecules that have homologous nucleotide sequences. Reciprocal exchanges can occur between any homologous donor and recipient sites. In E coli, the product of the recA gene is essential for generalized recombination, but other gene products also participate.

Site-specific recombination involves reciprocal exchanges only between specific sites in donor and recipient DNA molecules. The recA gene product is not required for site-specific recombination. Integration of the temperate bacteriophage l into the chromosome of E coli is a well-studied example of site-specific recombination. The specific attachment (att) sites on the E coli chromosome and l phage DNA have a common core sequence of 15 nucleotides, within which reciprocal recombination occurs, flanked by adjacent sequences that are not homologous in the phage and bacterial genomes. In phage l the product of the int gene (integrase) is required for the site-specific integration event in lysogenization; the products of the int and xis (excisionase) genes are both needed for the complementary site-specific excision event that occurs during induction of lytic phage development in lysogenic cells.

Illegitimate recombination is the term used to describe nonhomologous, aberrant recombination events such as those involved in formation of specialized transducing phages. The mechanisms of illegitimate recombination are unknown.

Transposons

Transposons are segments of DNA that can move from one site in a DNA molecule to other target sites in the same or a different DNA molecule. The process is called transposition and occurs by a mechanism that is independent of generalized recombination. Transposons are important genetic elements because they cause mutations, mediate genomic rearrangements, function as portable regions of genetic homology, and acquire new genes and contribute to their dissemination within bacterial populations. Insertion of a transposon often interrupts the linear sequence of a gene and inactivates it. Transposons have a major role in causing deletions, duplications, and inversions of DNA segments as well as fusions between replicons. Transposons are not self-replicating genetic elements, however, and they must integrate into other replicons to be maintained stably in bacterial genomes.

Most transposons share a number of common features. Each transposon encodes the functions necessary for its transposition, including a transposase enzyme that interacts with specific sequences at the ends of the transposon. During transposition a short sequence of target DNA is duplicated, and the transposon is inserted between the directly repeated target sequences. The length of this short duplication varies, but is characteristic for each transposon. The duplication is presumed to involve asymmetric cleavage of DNA at the target site, followed by synthesis of new complementary strands corresponding to the region between the cleavage sites. Some transposons insert into almost any target sequence, whereas others have relatively stringent target specificity. Two types of transposition are recognized.

Excision of the transposon from a donor site followed by its insertion into a target site is called nonreplicative transposition. If the transposon at a donor site is replicated and a copy is inserted into the target site, however, the process is called replicative transposition. The process of replicative transposition can involve formation of a cointegrate, a single circular DNA molecule consisting of two replicons joined with copies of the transposon in an alternating sequence. Resolution of the cointegrate into its component replicons is often accomplished by a transposon-encoded resolvase that catalyzes site-specific recombination between the transposons. Generalized recombination between homologous transposons can also lead to the formation or resolution of cointegrates. Transposition differs from site-specific recombination by duplicating a segment of the target sequence and by using a variety of different target sequences for a single donor sequence.

Most transposons in bacteria can be separated into three major classes. Insertion sequences and related composite transposons comprise the first class. Insertion sequences are simplest in structure and encode only the functions needed for transposition. The known insertion sequences vary in length from approximately 780 to 1500 nucleotide pairs, have short (15-25 base pair) inverted repeats at their ends, and are not closely related to each other. The DNA between the inverted terminal repeats contains one (or rarely two) transposase genes and does not encode a resolvase. Complex transposons vary in length from about 2,000 to more than 40,000 nucleotide pairs and contain insertion sequences (or closely related sequences) at each end, usually as inverted repeats. The entire complex element can transpose as a unit. The DNA between the terminal insertion sequences of complex transposons encodes multiple functions that are not essential for transposition. In medically important bacteria, genes that determine production of adherence antigens, toxins, or other virulence factors, or specify resistance to one or more antibiotics, are often located in complex transposons. Well-known examples of complex transposons are Tn5 and Tn10, which determine resistance to kanamycin and tetracycline, respectively. The complex transposons probably evolve by transposition of homologous insertion sequences to nearby sites within a DNA molecule.

The second class of transposons consists of the highly homologous TnA family. These transposons have longer (35 to 40 base pair) terminal inverted repeats than the complex transposons described above, but they lack terminal insertion sequences. All members of the family encode both transposase and resolvase functions. Well known examples from the TnA transposon family include the ampicillin resistance transposon Tn3 and Tn1000 (the gamma-delta transposon) found in the F plasmid. The TnA family has an important place in the history of medical microbiology. The development of high-level resistance to ampicillin in Haemophilus influenzae and Neisseria gonorrhoeae during the 1970s, which severely limited the usefulness of ampicillin for treatment of gonorrhea and Haemophilus infections in areas where such strains became prevalent, was caused by dissemination of ampicillin resistance determinants from TnA transposons in plasmids of the Enterobacteriaceae to plasmids in Haemophilus and Neisseria.

The third class of transposons consists of bacteriophage Mu and related temperate phages. The entire phage genome functions as a transposon, and replication of the phage DNA during vegetative growth occurs by replicative transposition. Prophage integration can occur at many different sites in the bacterial chromosome and often causes mutations. For that reason Mu and related phages are sometimes called mutator phages.

A fourth class of transposons, discovered in Gram-positive bacteria and represented by Tn917, consists of conjugative transposons that are completely different from the transposons described above. The conjugative transposon does not generate a duplication of the target sequence into which it inserts, and in Gram-positive bacteria the host strain carrying the transposon can act as a conjugal donor. Recipient bacteria need not be closely related to the donor bacterium. The transposon is excised from the chromosome of the donor and transmitted by conjugation to the recipient, where it integrates randomly into the chromosome. Tn917 encodes tetracycline resistance, but other larger conjugative transposons may encode additional antibiotic resistances. Conjugative transposons appear to be a major cause of the spread of antibiotic resistance in Gram-positive bacteria.

Some roles of transposons in bacterial evolution are illustrated by considering enteric Gram-negative bacteria and the structure of their plasmids. Bacteria collected during the pre-antibiotic era contained many plasmids, but they usually lacked resistance determinants. Many of the R plasmids from current clinical isolates belong to the same incompatibility groups as plasmids found previously, but they also determine resistance to multiple antibiotics. The close relationships between their replicons provide strong evidence that many current R plasmids evolved from the older plasmids by acquisition of resistance determinants. Some of the multiple antibiotic resistant plasmids have individual transposons with several resistance determinants, others have multiple resistance transposons located at separate sites, and still others contain complex hybrid resistance transposons formed by integration of one transposon into another.

The stepwise acquisition of resistance determinants can lead, in some cases, to the formation of composite transposons that encode multiple resistance determinants. Therapeutic use of antibiotics and their incorporation into animal feeds provide selective advantages for bacteria with R plasmids, whereas conjugation, transformation and transfection provide means for dissemination of R plasmids within and between bacterial species. After a plasmid carrying a transposon is introduced into a new bacterial host, the transposon and its determinants can jump into the chromosome or indigenous plasmids of the new host. Therefore, stability of the mobilizing plasmid in a new bacterial host is not essential for persistence of genetic determinants located on a transposon.

RECOMBINATION DNA AND GENE CLONING

Many methods are available to make hybrid DNA molecules in vitro (recombinant DNA) and to characterize them. Such methods include isolating specific genes in hybrid replicons, determining their nucleotide sequences, and creating mutations at designated locations (site-directed mutagenesis). A clone is a population of organisms or molecules derived by asexual reproduction from a single ancestor. Gene cloning is the process of incorporating foreign genes into hybrid DNA replicons. Cloned genes can be expressed in appropriate host cells, and the phenotypes that they determine can be analyzed. Some key concepts underlying representative methods are summarized here.

The first step in gene cloning is to make fragments of the donor DNA by mechanical or enzymatic methods. Certain restriction endonucleases, designated as class II, are particularly useful for preparing defined fragments of DNA molecules. They cleave both strands of double-stranded DNA molecules at specific, palindromic sequences (restriction sites) that usually vary from four to eight nucleotides in length, and the resulting DNA fragments are called restriction fragments. Some restriction endonucleases cleave at coincident sites to create blunt-ended DNA fragments, and others cut at staggered positions to create DNA fragments with short, self-complementary, single-stranded 5' or 3' ends.

The random probability that n adjacent nucleotides in a DNA strand will correspond to a specific restriction site is approximately 1/4n. Sites for enzymes that recognize unique 4, 6, or 8 nucleotide targets are likely to occur about once in every 256, 4096, or 65,536 nucleotides, respectively. By choosing appropriate restriction enzymes, specific DNA molecules, including bacterial chromosomes, plasmids, and phage genomes, can be digested into sets of restriction fragments that have appropriate sizes for specific applications.

A restriction map identifies the positions of target sites for specific restriction endonucleases in a DNA molecule. Restriction maps are available for many cloned DNA fragments, plasmids and phage genomes, as well as for the entire chromosome of E coli and several other bacteria.

The second step in gene cloning is to create hybrid replicons consisting of donor DNA fragments and a cloning vector. Cloning vectors are small plasmid or phage replicons that have one or more restriction sites into which foreign DNA can be inserted. Hybrid replicons are produced by using DNA ligase to join the restricted vector DNA with donor DNA fragments that have compatible ends, or, alternatively, synthetic oligonucleotides are used as linkers to create compatibility between donor and vector DNA molecules with different ends. Ligating a vector to a heterogeneous set of DNA fragments from a donor genome is called shotgun cloning, and the collection of recombinant DNA molecules that contains the various fragments is called a genomic library. If a specific DNA fragment is available, it can be incorporated into a recombinant replicon by direct cloning into an appropriate vector chosen from the wide variety of vectors available. Plasmid and phage vectors are used mainly to clone small inserts usually less than 10 kbp.

Examples of more special purpose vectors include cosmids, which are plasmid vectors that can be packaged into phage capsids (lambda cosmids accept inserts up to 30-40 kbp), and phagemids, which are plasmid-phage hybrid replicons that can exist either as plasmids or as single-stranded DNA phages under different experimental conditions. Phage P1 cosmids can accept inserts up to 100 kbp, and still larger DNA molecules can be cloned in yeast artificial chromosomes (YACs) which can stably maintain inserts up to and exceeding 1 Mbp in size. Other specialized vectors detect promoters, transcription termination signals, or other regulatory elements within foreign DNA inserts or, conversely, provide promoters from which transcription of cloned genes can be initiated.

The final steps in gene cloning are to introduce hybrid replicons into appropriate recipient cells and test them for expression of donor genes of interest. Prokaryotic cells (including bacteria) or eukaryotic cells (including yeast, animal or plant cells) can be used as recipients, but they differ with respect to their permissiveness for specific replicons, the transcriptional signals that they recognize, and the post-translational modifications of protein structure that they can accomplish. Recombinant DNA molecules produced in vitro can be introduced directly into recipient cells by transformation or transfection. In addition, clones in cosmid or phage vectors can be packaged into phage coats and introduced into susceptible recipient cells by transduction. By using specialized vectors (shuttle vectors) that can replicate in multiple cell types, genes from any organism can be cloned and manipulated in a convenient bacterial system and subsequently reintroduced into cells of the original organism for analysis in their natural environment.

Many methods are available to identify bacteria that contain recombinant DNA molecules. Most cloning vectors have genes for traits that can be positively selected, such as resistance to antibiotics. Furthermore, it is often possible to introduce foreign DNA into the cloning vector at a site that inactivates a nonessential, but easily recognizable, vector function. If both of these conditions are fulfilled, bacteria that contain recombinant molecules can be selected and distinguished easily from bacteria that contain only the vector. Bacteria in a genomic library that contain a particular cloned gene can be identified by using biochemical or immunologic methods to test for the desired gene product. Alternatively, the cloned gene of interest can be detected directly by using nucleic acid hybridization methods, provided that a specific DNA or RNA probe is available. Because insertion of foreign DNA into a cloning vector at an appropriate site does not inactivate its ability to replicate in appropriate recipient cells, hybrid replicons of interest can be amplified by replication, and the recombinant DNA molecules or their gene products can be purified and studied. The ability to purify specific DNA molecules made it feasible to develop enzymatic and chemical methods for determining their nucleotide sequences, and current methods for introducing mutations at defined sites in cloned genes are based on knowing their restriction maps or nucleotide sequences.

Recombinant DNA methods make it feasible to clone specific DNA fragments from any source into vectors that can be studied in well-characterized bacteria, in eukaryotic cells, or in vitro. Applications of DNA cloning are expanding rapidly in all fields of biology and medicine. In medical genetics such applications range from the prenatal diagnosis of inherited human diseases to the characterization of oncogenes and their roles in carcinogenesis. Pharmaceutical applications include large-scale production from cloned human genes of biologic products with therapeutic value, such as polypeptide hormones, interleukins, and enzymes. Applications in public health and laboratory medicine include development of vaccines to prevent specific infections and probes to diagnose specific infections by nucleic acid hybridization or polymerase chain reaction (PCR). The latter process uses oligonucleotide primers and DNA polymerase to amplify specific target DNA sequences during multiple cycles of synthesis in vitro, making it possible to detect rare target DNA sequences in clinical specimens with great sensitivity.

Regulation of Gene Expression

The phenotypic properties of bacteria are determined by their genotypes and growth conditions. For bacteria in pure culture, changes in growth conditions often result in predictable physiological adaptations in all members of the population. Typically, essential gene products are made in amounts that permit fastest growth in the given environment, and products required under special circumstances are made only when they are needed.

Physiological adaptations are often associated with changes in metabolic activities. The flow of metabolites through particular biochemical pathways can be controlled both by regulating the synthesis of specific enzymes and by altering the activities of existing enzymes. Mechanisms that regulate expression of genes by affecting synthesis of specific gene products are discussed here.

Specific regulation involves a gene or group of genes involved in a particular metabolic process. Induction and repression enable bacteria to regulate production of specific gene products in response to appropriate signals. Generally catabolic enzymes are induced when the substrate for the pathway is present in the growth medium, and biosynthetic enzymes are repressed by the product of the pathway. Enzymes that participate in a single biochemical pathway often occupy adjacent positions on the bacterial chromosome and are coordinately induced or repressed.

They form an operon, a group of contiguous genes that is transcribed as a single unit and translated to produce the corresponding gene products. Organization into an operon is an important strategy for coordinately regulating the expression of genes in bacteria.

Operons that can be induced or repressed are controlled by binding of specific regulatory proteins to particular nucleotide sequences that function as regulatory sites within the operon. Comparison of the amino acid sequences of many of these different regulatory proteins showed that they could be grouped together into families of regulators (e.g. the lysR family of proteins) that may have evolved from common ancestoral genes. Members of the lysR family include regulators of such diverse phenomena as lysine, cysteine and methionine metabolism in E coli and iron repression in V cholerae.

Global regulation simultaneously alters expression of a group of genes and operons, collectively called a regulon, that are controlled by the same regulatory signal. Global regulation determines responses of bacteria to basic nutrients such as carbon, nitrogen or phosphate, reactions to stresses such as DNA damage or heat shock, and synthesis by pathogens of specific virulence factors during growth in their host animals.

The amount of a specific protein in a bacterial cell can vary from none to many thousands of molecules. This wide range is often determined by the combined action of several regulatory mechanisms that affect expression of the corresponding structural gene. Regulation is achieved by determining how often a gene is transcribed into functional mRNA, how efficiently the mRNA is translated into protein, how rapidly the mRNA is degraded, how rapidly the protein product turns over, and whether the activity of the protein product can be altered by allosteric effects or covalent modifications.

mRNAs as Transcriptional Units

Gene expression begins with DNA-dependent RNA polymerase (RNA polymerase) catalyzing the transcription of specific mRNA from one strand of a DNA template. Binding of RNA polymerase to DNA occurs at specific sites called promoters, and transcription begins adjacent to the promoter. Strong promoters can interact efficiently with RNA polymerase and initiate transcription at a high rate; weak promoters initiate transcription at slow rates. In either case, mRNA is synthesized from its 5' end toward its 3' end at an approximately constant rate until the RNA polymerase recognizes another specific site called a terminator. RNA polymerase then dissociates from the template, and transcription of the mRNA is completed.

Individual mRNA molecules may code for one or more polypeptides. Transcription of an operon produces a polycistronic mRNA that codes for several polypeptides. Translation of polycistronic mRNAs leads to coordinate synthesis of the encoded polypeptides, but each polypeptide is synthesized as a separate molecule. A specific ribosome binding site is located just upstream from the start of each coding sequence on the mRNA molecule.

Messenger RNAs in bacteria are degraded rapidly with an average half life of several minutes, in contrast to tRNAs and rRNAs which are much more stable. Although mRNAs represent about half of the newly synthesized RNA, they represent only a small fraction of the total RNA. The short half-life of mRNAs has important consequences for gene expression. If the synthesis of a specific mRNA is prevented, production of the corresponding polypeptides declines rapidly.

Control of gene expression occurs by regulating one or more of the steps in the pathway from the DNA template to the active gene product. Simultaneous regulation at several levels permits greater control over gene expression than would be possible with a single regulatory mechanism. The most common way to regulate gene expression in bacteria is to control the production of specific mRNAs. Since the rate of elongation of an RNA molecule is approximately constant, the major factors that control mRNA synthesis are the rate of initiation and the probability that a full length transcript will be produced.

Regulation of Transcription Initiation

Some mRNAs in bacteria are synthesized at constant rates, resulting in constitutive production of the encoded polypeptides. The amounts of specific mRNAs and polypeptides produced from different constitutive genes vary greatly, however, and often reflect differences in strength of the promoters for those genes.

Transcription of many operons is regulated in response to changing environmental conditions. The promoters determine the maximum rate of transcription initiation for such operons, but regulatory proteins participate in controlling transcription. Nucleotide sequences in operons to which specific regulatory proteins bind are called regulatory sites or operators. Operators and promoters are located close together within operons and may have overlapping DNA sequences. The binding of regulatory proteins to operators can either increase (positive regulation) or decrease (negative regulation) the frequency of transcription initiation. Proteins that function as negative regulators are usually called repressors. Because regulatory proteins can diffuse through the cytoplasm, the structural genes for regulatory proteins do not have to be linked to the target operons.

The ability to sense the presence or absence of specific compounds and change the rates of synthesis of appropriate gene products are central to the control of gene expression. Regulatory proteins offer one solution to this problem of stimulus-response coupling. Many regulatory proteins are bifunctional and bind not only to appropriate operators but also to specific effectors, which are small molecules such as particular sugars, amino acids, and other metabolites. Furthermore, regulatory proteins are allosteric, meaning that they can exist in different conformations which exhibit different binding affinities for their cognate operators and effectors. A sufficient concentration of effector favors formation of the regulatory protein-effector complex, which has either high or low affinity for the operator in any specific case. In negatively regulated systems the effector functions as a corepressor if the regulatory protein-effector complex is the active repressor, and the effector functions as an inducer and causes derepression if the free regulatory protein is the active repressor. Conversely, in positively regulated systems, the effector stimulates expression of the operon if the regulatory protein-effector complex is the positive regulator, and the effector inhibits expression of the operon if the free regulatory protein is the positive regulator.

The lactose (lac) operon of E coli is an example of an inducible, negatively regulated operon. The lacI gene codes for a repressor that binds to the lac operator and prevents transcription from the lac promoter. The structural gene for this repressor is separate from the lac operon, and the repressor is synthesized constitutively at a low rate. When inducer binds to the lac repressor, the complex cannot bind to the operator and cannot prevent binding of the RNA polymerase to the promoter. If other conditions are favorable, the lac operon is expressed, resulting in synthesis of b-galactosidase, b-galactoside permease and b-galactoside transacetylase. The lac operon can be induced by lactose or by structurally related compounds such as isopropyl-b-D-thiogalactoside (IPTG). IPTG is called a gratuitous inducer because it induces the lac operon, but is not a substrate for b-galactosidase. Negative regulation also occurs in many biosynthetic operons in E coli. In such operons a product of the biosynthetic pathway functions as the effector for the negative regulatory system.

The arabinose (ara) operon in E coli is both positively and negatively regulated. In the presence of arabinose the regulatory protein stimulates transcription of the ara operon. In the absence of arabinose, however, the regulatory protein represses the ara operon.

Operons are often controlled by more than one mechanism. When E coli is grown in a medium containing glucose and an alternative carbon source such as lactose or arabinose, induction of the lac or ara operon and utilization of the lactose or arabinose are delayed until the glucose has been consumed. This phenomenon is called diauxic growth. The failure to induce the lac or ara operon in the presence of glucose is an example of catabolite repression. The lac and ara operons are positively regulated by cyclic-3',5'-adenosine monophosphate (cAMP) and the catabolite gene activator (CAP) protein (the product of the crp gene). The cAMP-CAP complex interacts with CAP binding sites in the regulatory regions of some operons, including the lac and ara operons, and stimulates transcription from the corresponding promoters. The level of intracellular cAMP in E coli is high during growth in the absence of glucose, and low during growth in the presence of glucose. Catabolite repression is due, therefore, to lack of activation of cAMP-dependent operons when the bacteria are grown in the presence of glucose or certain other rapidly metabolizable carbon sources.

REGULATION OF TRANSCRIPTION TERMINATION

Attenuation is a mechanism for regulating operons by terminating transcription of mRNA prematurely. Attenuation is common in biosynthetic operons, including the trp, histidine (his), threonine (thr), isoleucine-valine (ilv), and phenylalanine (phe) operons. The trp operon in E coli is controlled both by repression and attenuation. In the presence of excess tryptophan, initiation of transcription from the trp promoter is repressed. In addition, however, those transcripts that are initiated from the trp promoter are usually terminated before any of the structural genes of the trp operon are transcribed. The concentration of intracellular tryptophan required to maintain repression exceeds that needed for attenuation. Such dual control enables the cell to fine tune the expression of the trp operon in response to decreasing concentrations of tryptophan.

The secondary structure of mRNA has an important role in the mechanism of attenuation. All mRNAs have a leader sequence between the transcriptional start site and the beginning of the coding sequence for the first structural gene. For amino acid biosynthetic operons that are subject to attenuation, the mRNA leader sequence has two distinctive features. It encodes a short peptide containing the amino-acid produced by the regulated pathway, and it can form alternative, mutually incompatible, double-stranded RNA structures that participate in regulatory events. For example, the peptide encoded by the trp mRNA leader sequence contains two adjacent tryptophan residues, and the peptide encoded by the his mRNA leader sequence has a series of seven consecutive histidine residues. Fig. 5-14 shows the trp operon and illustrates alternative secondary structures in the leader sequence of trp mRNA. There are three possible secondary structures for this region, called the pause site (segments 1+2), the anti-terminator (segments 2+3), and the attenuator (segments 3+4). Segment 1 of the pause site overlaps with the coding region for the trpL peptide.

Which secondary structures are formed depends on efficiency of translation of the trpL peptide. When segments 1 and 2 are transcribed, they immediately anneal and cause the RNA polymerase to pause temporarily. Subsequent initiation of translation of the trpL peptide disrupts the pause site and allows RNA polymerase to continue transcription. If tryptophan is present, transcription of segments 3 and 4 and formation of the attenuator structure occurs while the ribosome is blocking segment 2, causing the RNA polymerase to terminate transcription. If tryptophan is deficient, however, tryptophanyl-tRNA is also deficient, and the ribosome stalls at the tryptophan codons in segment 1. This allows segment 2 to anneal with newly synthesized segment 3 to form the antiterminator, thereby making segment 3 unavailable to anneal with segment 4. Formation of the attenuator is therefore prevented, and the RNA polymerase transcribes the entire trp operon. In this manner depletion of tryptophan (actually the supply of tryptophanyl-tRNA) is coupled to regulation of transcription of the biosynthetic operon for tryptophan.

In E coli transcription and translation are functionally coupled. Nonsense mutations that cause premature termination of translation often cause decreased transcription of more distal genes in the same operon. This phenomenon is called polarity.

Ribosomes usually initiate translation of a growing mRNA molecule prior to completion of transcription, and such translation masks sites that would otherwise cause the RNA polymerase to terminate transcription. Premature termination of translation by a nonsense codon dissociates the ribosomes from the mRNA and enables RNA polymerase to interact with the unmasked transcription termination sites. In some biological systems, including phage lambda, antitermination is used as a positive regulatory mechanism to control gene expression. Immediately after infection of E coli by lambda, RNA polymerase binds to two promoters in lambda DNA and initiates divergent primary transcripts which terminate at specific sites on the lambda genome. A protein encoded by one of the primary transcripts interacts with RNA polymerase and enables it to continue transcription through the primary termination sites, thereby expressing a second set of lambda genes. One of the products encoded by a secondary transcript blocks termination of another mRNA and activates expression of a third set of genes. Antitermination has a key role, therefore, in controlling the cascade of gene expression during lytic growth of phage lambda. Antitermination is also involved in the regulation of E coli rRNA operons.

Regulation of Translation

The ribosome binding site on mRNA is complementary to a sequence at the 3' end of 16S rRNA. Interaction between these sequences facilitates formation of the initiation complex for protein synthesis. Both the extent of homology with 16S rRNA and the spacing of the ribosome binding site from the initiation codon affect the efficiency of translation initiation. Codon usage in mRNA also influences translation efficiency. Messenger RNAs for proteins that are required in large amounts tend to use codons that are translated by the most abundant species of tRNA, and the converse is also true.

Translational control is important for regulation of synthesis of ribosomal proteins. Production of ribosomes involves a high metabolic cost for bacteria, and at high growth rates ribosomes can constitute nearly one-half of the cell weight. Most ribosomal proteins and rRNAs are found assembled into ribosomes, and the pool of free ribosomal subunits is very small. The genes for ribosomal proteins are organized into several operons. Certain of the free ribosomal proteins directly inhibit the translation of the polycistronic mRNAs that encode them, thereby ensuring that synthesis of ribosomal proteins is balanced with the requirement for their utilization.

REGULONS AND SIGNAL TRANSDUCING PROTEINS

A regulon is a group of genes or operons controlled by a common regulator. There are several advantages to placing different operons under control by the same regulator. It enables the sensing of a single stimulus to be coupled to expression of a large number of genes that may be needed for an appropriate response, and it eliminates the requirement for the coordinately regulated genes to be linked on the bacterial chromosome. The stimulus to which the regulon responds can be an intracellular component or an environmental signal. Individual operons may also be subject to regulation by several different mechanisms and expressed under conditions that differ from those affecting the whole regulon.

More than 40 different regulons have been identified in E coli. Specific examples of regulons that respond to intracellular components include the cAMP-CAP regulon described previously and the regulons controlled by the stringent response and the SOS response. When ribosomes encounter uncharged tRNA molecules during protein synthesis, the stringent response is activated and results in prompt cessation of rRNA synthesis. A novel nucleotide called guanosine-3'-diphosphate-5'-diphosphate (ppGpp) accumulates during amino-acid starvation. The ppGpp produced by idling ribosomes appears to be a mediator of the stringent response, but the precise mechanism causing inhibition of rRNA synthesis is unknown. The SOS response is associated with damage to DNA and involves induction of more than 20 genes involved in several DNA repair pathways. The product of the recA gene detects inhibition of DNA synthesis and initiates events leading to proteolytic cleavage and inactivation of the repressor for the SOS pathway, encoded by the lexA gene.

Some regulons are induced by specific environmental stimuli, such as nutrient limitation or osmotic stress. Often operons from more than one regulon may be induced, and the term stimulon has been used to describe the set of genes so induced. Typically, bacteria sense such environmental conditions by two component systems. The first component is a membrane-spanning protein with extracellular and intracellular domains. Its extracellular domain detects the environmental stimulus, and its cytoplasmic domain transmits the signal. The second component is a bifunctional cytoplasmic protein. It has a receiver domain that interacts with the transmitter module of the first component, as well as an effector domain that controls expression of the corresponding regulon. The transmitter and receiver modules of the two component regulatory systems from a wide variety of regulons are genetically related and share amino-acid homology. The signal-detecting and effector domains of the proteins from different regulons vary, however, and determine the signal that is detected and the operons that are activated or repressed in response to that signal.

Global regulation has an important role in the physiology of pathogenic bacteria. For example, Vibrio cholerae and Bordetella pertussis express many of their virulence determinants under the control of signal transducing systems that are related to the two component systems described above. The expression of proteins needed for the invasive phenotype is controlled by temperature in Shigella. Yersinia enterocolitica senses both the environmental temperature and the concentration of calcium ions and couples these signals to the expression of genes and cellular location of the gene products that are appropriate for an intracellular or extracellular environment. In host tissues the concentration of free iron is extremely low, and most pathogenic bacteria have high affinity iron transport systems that are induced under low-iron conditions. The synthesis of diphtheria toxin by C diphtheriae, Shiga toxin by Shigella dysenteriae, exotoxin A by Pseudomonas aeruginosa, and other specific proteins in many pathogenic bacteria is induced under conditions of iron-limited growth. These examples illustrate how environmental factors can regulate the expression of virulence genes in pathogenic bacteria.

5

Normal Flora

INTRODUCTION

A diverse microbial flora is associated with the skin and mucous membranes of every human being from shortly after birth until death. The human body, which contains about 10^{13} cells, routinely harbors about 10^{14} bacteria. This bacterial population constitutes the normal microbial flora. The normal microbial flora is relatively stable, with specific genera populating various body regions during particular periods in an individual's life. Microorganisms of the normal flora may aid the host (by competing for microenvironments more effectively than such pathogens as Salmonella spp or by producing nutrients the host can use), may harm the host (by causing dental caries, abscesses, or other infectious diseases), or may exist as commensals (inhabiting the host for long periods without causing detectable harm or benefit). Even though most elements of the normal microbial flora inhabiting the human skin, nails, eyes, oropharynx, genitalia, and gastrointestinal tract are harmless in healthy individuals, these organisms frequently cause disease in compromised hosts. Viruses and parasites are not considered members of the normal microbial flora by most investigators because they are not commensals and do not aid the host.

Significance of the Normal Flora

The fact that the normal flora substantially influences the well-being of the host was not well understood until germ-free animals became available. Germ-free animals were obtained by cesarean section and maintained in special isolators; this allowed the investigator to raise them in an environment free from detectable viruses, bacteria, and other organisms. Two interesting observations were made about animals raised under germ-free conditions. First, the germ-free animals lived almost twice as long as their conventionally maintained counterparts, and second, the major causes of death were different in the two groups. Infection often caused death in conventional animals, but intestinal atonia frequently killed germ-free animals. Other investigations showed that germ-free animals have anatomic, physiologic, and immunologic features not shared with conventional animals. For example, in germ-free animals, the alimentary lamina propria is underdeveloped, little or no immunoglobulin is present in sera or secretions, intestinal motility is reduced, and the intestinal epithelial cell renewal rate is approximately one-half that of normal animals (4 rather than 2 days).

Although the foregoing indicates that bacterial flora may be undesirable, studies with antibiotic treated animals suggest that the flora protects individuals from pathogens. Investigators have used streptomycin to reduce the normal flora and have then infected animals with streptomycin-resistant Salmonella. Normally, about 10^6 organisms are needed to establish a gastrointestinal infection, but in streptomycin-treated animals whose flora is altered, fewer than 10 organisms were needed to cause infectious disease. Further studies suggested that fermentation products (acetic and butyric acids) produced by the normal flora inhibited Salmonella growth in the gastrointestinal tract.

The normal flora in humans usually develops in an orderly sequence, or succession, after birth, leading to the stable populations of bacteria that make up the normal adult flora. The main factor determining the composition of the normal flora in a body region is the nature of the local environment, which is determined by pH, temperature, redox potential, and oxygen, water, and nutrient levels. Other factors such as peristalsis, saliva, lysozyme secretion, and secretion of immunoglobulins also play roles in flora control. The local environment is like a concerto in which one principal instrument usually dominates. For example, an infant begins to contact organisms as it moves through the birth canal. A Gram-positive population (bifidobacteria arid lactobacilli) predominates in the gastrointestinal tract early in life if the infant is breast-fed. This bacterial population is reduced and displaced somewhat by a Gram-negative flora (Enterobacteriaceae) when the baby is bottle-fed. The type of liquid diet provided to the infant is the principal instrument of this flora control; immunoglobulins and, perhaps, other elements in breast milk may also be important.

What, then, is the significance of the normal flora? Animal and some human studies suggest that the flora influences human anatomy, physiology, lifespan, and, ultimately, cause of death. Although the causal relationship of flora to death and disease in humans is accepted, of her roles of the human microflora need further study.

NORMAL FLORA OF SKIN

Skin provides good examples of various microenvironments. Skin regions have been compared to geographic regions of Earth: the desert of the forearm, the cool woods of the scalp, and the tropical forest of the armpit. The composition of the dermal microflora varies from site to site according to the character of the microenvironment. A different bacterial flora characterizes each of three regions of skin: (1) axilla, perineum, and toe webs; (2) hand, face and trunk; and (3) upper arms and legs. Skin sites with partial occlusion (axilla, perineum, and toe webs) harbor more microorganisms than do less occluded areas (legs, arms, and trunk). These quantitative differences may relate to increased amount of moisture, higher body temperature, and greater concentrations of skin surface lipids. The axilla, perineum, and toe webs are more frequently colonized by Gram-negative bacilli than are drier areas of the skin.

The number of bacteria on an individual's skin remains relatively constant; bacterial survival and the extent of colonization probably depend partly on the exposure of skin to a particular environment and partly on the innate and species-specific bactericidal activity in skin. Also, a high degree of specificity is involved in the adherence of bacteria to epithelial surfaces. Not all bacteria attach to skin; staphylococci, which are the major element of the nasal flora, possess a distinct advantage over viridans streptococci in colonizing the nasal mucosa. Conversely, viridans streptococci are not seen in large numbers on the skin or in the nose but dominate the oral flora.

The microbiology literature is inconsistent about the density of bacteria on the skin; one reason for this is the variety of methods used to collect skin bacteria. The scrub method yields the highest and most accurate counts for a given skin area. Most microorganisms live in the superficial layers of the stratum corneum and in the upper parts of the hair follicles. Some bacteria, however, reside in the deeper areas of the hair follicles and are beyond the reach of ordinary disinfection procedures. These bacteria are a reservoir for recolonization after the surface bacteria are removed.

Staphylococcus epidermidis

S epidermidis is a major inhabitant of the skin, and in some areas it makes up more than 90 percent of the resident aerobic flora.

Staphylococcus aureus

The nose and perineum are the most common sites for S aureus colonization, which is present in 10 percent to more than 40 percent of normal adults. S aureus is prevalent (67 percent) on vulvar skin. Its occurrence in the nasal passages varies with age, being greater in the newborn, less in adults. S aureus is extremely common (80 to 100 percent) on the skin of patients with certain dermatologic diseases such as atopic dermatitis, but the reason for this finding is unclear.

Micrococci

Micrococci are not as common as staphylococci and diphtheroids; however, they are frequently present on normal skin. Micrococcus luteus, the predominant species, usually accounts for 20 to 80 percent of the micrococci isolated from the skin.

DIPHTHEROIDS (CORYNEFORMS)

The term diphtheroid denotes a wide range of bacteria belonging to the genus Corynebacterium. Classification of diphtheroids remains unsatisfactory; for convenience, cutaneous diphtheroids have been categorized into the following four groups: lipophilic or nonlipophilic diphtheroids; anaerobic diphtheroids; diphtheroids producing porphyrins (coral red fluorescence when viewed under ultraviolet light); and those that possess some keratinolytic enzymes and are associated with trichomycosis axillaris (infection of axillary hair). Lipophilic diphtheroids are extremely common in the axilla, whereas nonlipophilic strains are found more commonly on glabrous skin. Anaerobic diphtheroids are most common in areas rich in sebaceous glands. Although the name Corynebacterium acnes was originally used to describe skin anaerobic diphtheroids, these are now classified as Propionibacterium acnes and as P granulosum. P acnes is seen eight times more frequently than P granulosum in acne lesions and is probably involved in acne pathogenesis. Children younger than 10 years are rarely colonized with P acnes. The appearance of this organism on the skin is probably related to the onset of secretion of sebum (a semi-fluid substance composed of fatty acids and epithelial debris secreted from sebaceous glands) at puberty. P avidum, the third species of cutaneous anaerobic diphtheroids, is rare in acne lesions and is more often isolated from the axilla.

Streptococci

Streptococci, especially ß-hemolytic streptococci, are rarely seen on normal skin. The paucity of ß-hemolytic streptococci on the skin is attributed at least in part to the presence of lipids on the skin, as these lipids are lethal to streptococci. Other groups of streptococci, such as a-hemolytic streptococci, exist primarily in the mouth, from where they may, in rare instances, spread to the skin.

Gram-Negative Bacilli

Gram-negative bacteria make up a small proportion of the skin flora. In view of their extraordinary numbers in the gut and in the natural environment, their scarcity on skin is striking. They are seen in moist intertriginous areas, such as the toe webs and axilla, and not on dry skin. Desiccation is the major factor preventing the multiplication of Gram-negative bacteria on intact skin. Enterobacter, Klebsiella, Escherichia coli, and Proteus spp are the predominant Gram-negative organisms found on the skin. Acinetobacter spp also occurs on the skin of normal individuals and, like other Gram-negative bacteria, is more common in the moist intertriginous areas.

Nail Flora

The microbiology of a normal nail is generally similar to that of the skin. Dust particles and other extraneous materials may get trapped under the nail, depending on what the nail contacts. In addition to resident skin flora, these dust particles may carry fungi and bacilli. Aspergillus, Penicillium, Cladosporium, and Mucor are the major types of fungi found under the nails.

Oral and Upper Respiratory Tract Flora

The oral flora is involved in dental caries and periodontal disease, which affect about 80 percent. of the population in the Western world. The oral flora, its interactions with the host, and its response to environmental factors are thoroughly discussed in another Chapter. Anaerobes in the oral flora are responsible for many of the brain, face, and lung infections that are frequently manifested by abscess formation.

The pharynx and trachea contain primarily those bacterial genera found in the normal oral cavity (for example, alpha-and ß-hemolytic streptococci); however, anaerobes, staphylococci, neisseriae, diphtheroids, and others are also present. Potentially pathogenic organisms such as Haemophilus, mycoplasmas, and pneumococci may also be found in the pharynx. Anaerobic organisms also are reported frequently. The upper respiratory tract is so often the site of initial colonization by pathogens (Neisseria meningitides, C diphtheriae, Bordetella pertussis, and many others) and could be considered the first region of attack for such organisms. In contrast, the lower respiratory tract (small bronchi and alveoli) is usually sterile, because particles the size of bacteria do not readily reach it. If bacteria do reach these regions, they encounter host defense mechanisms, such as alveolar macrophages, that are not present in the pharynx.

GASTROINTESTINAL TRACT FLORA

The stomach is a relatively hostile environment for bacteria. It contains bacteria swallowed with the food and those dislodged from the mouth. Acidity lowers the bacterial count, which is highest (approximately 10^3 to 10^6 organisms/g of contents) after meals and lowest (frequently undetectable) after digestion. Some Helicobacter species can colonize the stomach and are associated with type B gastritis and peptic ulcer disease. Aspirates of duodenal or jejunal fluid contain approximately 10^3 organisms/ml in most individuals. Most of the bacteria cultured (streptococci, lactobacilli, Bacteroides) are thought to be transients. Levels of 10^5 to about 10^7 bacteria/ml in such aspirates usually indicate an abnormality in the digestive system (for example, achlorhydria or malabsorption syndrome). Rapid peristalsis and the presence of bile may explain in part the paucity of organisms in the upper gastrointestinal tract. Further along the jejunum and into the ileum, bacterial populations begin to increase, and at the ileocecal junction they reach levels of 10^6 to 10^8 organisms/ml, with streptococci, lactobacilli, Bacteroides, and bifidobacteria predominating.

Concentrations of 10^9 to 10^{11} bacteria/g of contents are frequently found in human colon and feces. This flora includes a bewildering array of bacteria (more than 400 species have been identified); nonetheless, 95 to 99 percent belong to anaerobic genera such as Bacteroides, Bifidobacterium, Eubacterium, Peptostreptococcus, and Clostridium. In this highly anaerobic region of the intestine, these genera proliferate, occupy most available niches, and produce metabolic waste products such as acetic, butyric, and lactic acids. The strict anaerobic conditions, physical exclusion (as is shown in many animal studies), and bacterial waste products are factors that inhibit the growth of other bacteria in the large bowel.

Although the normal flora can inhibit pathogens, many of its members can produce disease in humans. Anaerobes in the intestinal tract are the primary agents of intra-abdominal abscesses and peritonitis. Bowel perforations produced by appendicitis, cancer, infarction, surgery, or gunshot wounds almost always seed the peritoneal cavity and adjacent organs with the normal flora. Anaerobes can also cause problems within the gastrointestinal lumen. Treatment with antibiotics may allow certain anaerobic species to become predominant and cause disease. For example, *Clostridium difficile,* which can remain viable in a patient undergoing antimicrobial therapy, may produce pseudomembranous colitis. Other intestinal pathologic conditions or surgery can cause bacterial overgrowth in the upper small intestine. Anaerobic bacteria can then deconjugate bile acids in this region and bind available vitamin B12 so that the vitamin and fats are malabsorbed. In these situations, the patient usually has been compromised in some way; therefore, the infection caused by the normal intestinal flora is secondary to another problem.

More information is available on the animal than the human microflora. Research on animals has revealed that unusual filamentous microorganisms attach to ileal epithelial cells and modify host membranes with few or no harmful effects. Microorganisms have been observed in thick layers on gastrointestinal surfaces and in the crypts of Lieberkuhn. Other studies indicate that the immune response can be modulated by the intestinal flora. Studies of the role of the intestinal flora in biosynthesis of vitamin K and other host-utilizable products, conversion of bile acids (perhaps to cocarcinogens), and ammonia production (which can play a role in hepatic coma) show the dual role of the microbial flora in influencing the health of the host. More basic studies of the human bowel flora are necessary to define their effect on humans.

Urogenital Flora

The type of bacterial flora found in the vagina depends on the age, pH, and hormonal levels of the host. Lactobacillus spp predominate in female infants (vaginal pH, approximately 5) during the first month of life. Glycogen secretion seems to cease from about I month of age to puberty. During this time, diphtheroids, *S epidermidis*, streptococci, and *E coli* predominate at a higher pH (approximately pH 7). At puberty, glycogen secretion resumes, the pH drops, and women acquire an adult flora in which L acidophilus, corynebacteria, peptostreptococci, staphylococci, streptococci, and Bacteroides predominate. After menopause, pH again rises, less glycogen is secreted, and the flora returns to that found in prepubescent females. Yeasts (Torulopsis and Candida) are occasionally found in the vagina (10 to 30 percent of women); these sometimes increase and cause vaginitis.

In the anterior urethra of humans, *S epidermidis*, enterococci, and diphtheroids are found frequently; *E coli*, Proteus, and Neisseria (nonpathogenic species) are reported occasionally (10 to 30 percent). Because of the normal flora residing in the urethra, care must be taken in clinically interpreting urine cultures; urine samples may contain these organisms at a level of 10^4/ml if a midstream (clean-catch) specimen is not obtained.

Conjunctival Flora

The conjunctival flora is sparse. Approximately 17 to 49 percent of culture samples are negative. Lysozyme, secreted in tears, may play a role in controlling the bacteria by interfering with their cell wall formation. When positive samples show bacteria, corynebacteria, neisseriae, and moraxellae are cultured. Staphylococci and streptococci are also present, and recent reports indicate that Haemophilus parainfluenzae is present in 25 percent of conjunctival samples.

Host Infection by Elements of the Normal Flora

This chapter has briefly described the normal human flora; however, the pathogenic mechanisms of various genera or the clinical syndromes in which they are involved was not discussed. Although such material is presented in other chapters, note that a breach in mucosal surfaces often results in infection of the host by members of the normal flora. Caries, periodontal disease, abscesses, foul-smelling discharges, and endocarditis are hallmarks of infections with members of the normal human flora. In addition, impairment of the host (for example, those with heart failure or leukemia) or host defenses (due to immunosuppression, chemotherapy, or irradiation) may result in failure of the normal flora to suppress transient pathogens or may cause members of the normal flora to invade the host themselves. In either situation, the host may die.

BACTERIAL PATHOGENESIS

Infection is the invasion of the host by microorganisms, which then multiply in close association with the host's tissues. Infection is distinguished from disease, a morbid process that does not necessarily involve infection (diabetes, for example, is a disease with no known causative agent). Bacteria can cause a multitude of different infections, ranging in severity from inapparent to fulminating.

The capacity of a bacterium to cause disease reflects its relative pathogenicity. On this basis, bacteria can be organized into three major groups. When isolated from a patient, frank or primary pathogens are considered to be probable agents of disease (e.g., when the cause of diarrheal disease is identified by the laboratory isolation of *Salmonella* spp from feces). Opportunistic pathogens are those isolated from patients whose host defense mechanisms have been compromised. They may be the agents of disease (e.g., in patients who have been predisposed to urinary tract infections with *Escherichia coli* by catheterization). Finally, some bacteria, such as Lactobacillus acidophilus, are considered to be nonpathogens, because they rarely or never cause human disease. Their categorization as nonpathogens may change, however, because of the adaptability of bacteria and the detrimental effect of modern radiation therapy, chemotherapy, and immunotherapy on resistance mechanisms. In fact, some bacteria previously considered to be nonpathogens are now known to cause disease. Serratia marcescens, for example, is a common soil bacterium that causes pneumonia, urinary tract infections, and bacteremia in compromised hosts.

Virulence is the measure of the pathogenicity of an organism. The degree of virulence is related directly to the ability of the organism to cause disease despite host resistance mechanisms; it is affected by numerous variables such as the number of infecting bacteria, route of entry into the body, specific and nonspecific host defense mechanisms, and virulence factors of the bacterium. Virulence can be measured experimentally by determining the number of bacteria required to cause animal death, illness, or lesions in a defined period after the bacteria are administered by a designated route. Consequently, calculations of a lethal dose affecting 50 percent of a population of animals (LD_{50}) or an effective dose causing a disease symptom in 50 percent of a population of animals (ED_{50}) are useful in comparing the relative virulence of different bacteria.

Pathogenesis refers both to the mechanism of infection and to the mechanism by which disease develops. The purpose of this chapter is to provide an overview of the many bacterial virulence factors and, where possible, to indicate how they interact with host defense mechanisms and to describe their role in the pathogenesis of disease. It should be understood that the pathogenic mechanisms of many bacterial diseases are poorly understood, while those of others have been probed at the molecular level. The relative importance of an infectious disease to the health of humans and animals does not always coincide with the depth of our understanding of its pathogenesis. This information is best acquired by reading each of the ensuing chapters on specific bacterial diseases, infectious disease texts, and public health bulletins.

Host Susceptibility

Susceptibility to bacterial infections depends on the physiologic and immunologic condition of the host and on the virulence of the bacteria. Before increased amounts of specific antibodies or T cells are formed in response to invading bacterial pathogens, the "nonspecific" mechanisms of host resistance (such as polymorphonuclear neutrophils and macrophage clearance) must defend the host against the microbes.

Development of effective specific immunity (such as an antibody response to the bacterium) may require several weeks. The normal bacterial flora of the skin and mucosal surfaces also serves to protect the host against colonization by bacterial pathogens. In most healthy individuals, bacteria from the normal flora that occasionally penetrate the body (e.g., during tooth extraction or routine brushing of teeth) are cleared by the host's cellular and humoral mechanisms. In contrast, individuals with defective immune responses are prone to frequent, recurrent infections with even the least virulent bacteria.

The best-known example of such susceptibility is acquired immune deficiency syndrome (AIDS), in which the $CD4^+$ helper lymphocytes are progressively decimated by human immunodeficiency virus (HIV). However, resistance mechanisms can be altered by many other processes. For example, aging often weakens both nonspecific and specific defense systems so that they can no longer effectively combat the challenge of bacteria from the environment. Infants are also especially susceptible to certain pathogens (such as group B streptococci because their immune systems are not yet fully developed and cannot mount a protective immune response to important bacterial antigens. In addition, some individuals have genetic defects of the complement system or cellular defenses (e.g., inability of polymorphonuclear neutrophils to kill bacteria). Finally, a patient may develop granulocytopenia as a result of a predisposing disease, such as cancer, or immunosuppressive chemotherapy for organ transplants or cancer.

Host resistance can be compromised by trauma and by some underlying diseases. An individual becomes susceptible to infection with a variety of bacteria if the skin or mucosa is breached, particularly in the case of severe wounds such as burns or contaminated surgical wounds. Cystic fibrosis patients, who have poor ciliary function and consequently cannot clear mucus efficiently from the respiratory tract, are abnormally susceptible to infection with mucoid strains of *Pseudomonas aeruginosa*, resulting in serious respiratory distress. Ascending urinary tract infections with *Escherichia coli* are common in women and are particularly troublesome in patients with urinary tract obstructions. A variety of routine medical procedures, such as tracheal intubation and catheterization of blood vessels and the urethra, increase the risk of bacterial infection. The plastic devices used in these procedures are readily colonized by bacteria from the skin, which migrate along the outside of the tube to infect deeper tissues or enter the bloodstream. Because of this problem, it is standard practice to change catheters frequently (e.g., every 72 hours for peripheral intravenous catheters).

Many drugs have been developed to treat bacterial infections. Antimicrobial agents are most effective, however, when the infection is also being fought by healthy phagocytic and immune defenses. Some reasons for this situation are the poor diffusion of antibiotics into certain sites (such as the prostate gland), the ability of many bacteria to multiply or survive inside cells (where many antimicrobial agents have little or no effect), the bacteriostatic rather than bactericidal action of some drugs, and the capacity of some organisms to develop resistance to multiple antibiotics.

Many bacterial pathogens are transmitted to the host by a vector, usually an arthropod. For example, Rocky Mountain spotted fever and Lyme disease are both vectored by ticks, and bubonic plague is spread by fleas. Susceptibility to these diseases depends partly on the host's contact with the vector.

PATHOGENIC MECHANISMS

Bacterial Infectivity

Factors that are produced by a microorganism and evoke disease are called virulence factors. Examples are toxins, surface coats that inhibit phagocytosis, and surface receptors that bind to host cells. Most frank (as opposed to opportunistic) bacterial pathogens have evolved specific virulence factors that allow them to multiply in their host or vector without being killed or expelled by the host's defenses. Many virulence factors are produced only by specific virulent strains of a microorganism. For example, only certain strains of *E coli* secrete diarrhea-causing enterotoxins.

Virulence factors should never be considered independently of the host's defenses; the clinical course of a disease often depends on the interaction of virulence factors with the host's response. An infection begins when the balance between bacterial pathogenicity and host resistance is upset. In essence, we live in an environment that favors the microbe, simply because the growth rate of bacteria far exceeds that of most eukaryotic cells. Furthermore, bacteria are much more versatile than eukaryotic cells in substrate utilization and biosynthesis. The high mutation rate of bacteria combined with their short generation time results in rapid selection of the best-adapted strains and species. In general, bacteria are much more resistant to toxic components in the environment than eukaryotes, particularly when the major barriers of eukaryotes (skin and mucous membranes) are breached.

From a practical standpoint, bacteria can be said to have a single objective: to multiply. Only a few of the vast number of bacterial species in the environment consistently cause disease in a given host. From a teleologic standpoint, it is not in the best interest of the pathogen to kill the host, because in most cases the death of the host means the death of the pathogen. The most highly evolved or adapted pathogens are the ones that acquire the necessary nutritional substances for growth and dissemination with the smallest expenditure of energy and least damage to the host. For example, *Rickettsia akari*, the etiologic agent of rickettsialpox, causes a mild, self-limited infection consisting of headache, fever, and a papulovesicular rash. Other members of the rickettsial group, such as *R rickettsii*, the agent of Rocky Mountain spotted fever, elicit more severe, life-threatening infections. Some bacteria that are poorly adapted to the host synthesize virulence factors (e.g., tetanus and diphtheria toxin) so potent that they threaten the life of the host.

Host Resistance

Although easily damaged, the skin represents one of the most important barriers of the body to the microbial world, which contains a diverse array of bacteria in enormous numbers.

Fortunately, most bacteria in the environment are relatively benign to individuals with normal immune systems. However, patients who are immunosuppressed, such as individuals receiving cancer chemotherapy or have AIDS, opportunistic microbial pathogens can establish life-threatening infections. Normally, microbes in the environment are prevented from entering the body by the skin and mucous membranes. The outermost surface of the skin consists of squamous cell epithelium, largely comprised of dead cells that are sloughed off as new cells are formed below them.

In addition to the skin barrier, mucous membranes of the respiratory, gastrointestinal, and urogenital systems represent other portals through which bacteria can gain access to the body. Like the squamous epithelial cells of the skin, the mucosal epithelial cells divide rapidly, and as the cells mature, they are pushed laterally toward the intestinal lumen and shed. The entire process is reported to require only 36-48 hours for complete replacement of the epithelium, which diminishes the number of bacteria associated with the epithelium. The skin surface is a dry, acidic environment, and the temperature is less than 37° C. The pores and crevices of the skin also are colonized by the "normal bacterial flora", which ensure competition for pathogens to which the skin is exposed. Similarly, the mucous layer that covers the epithelia contains hostile substances to microbial colonization. Protective levels of lysozyme, lactoferrin, and lactoperoxidase in the mucus either kill bacteria or restrict their growth. In addition, the mucus contains secretory immunoglobulins (predominantly sIgA) synthesized by plasma cells resident in the submucosal tissue During the normal course of life, individuals develop local antibodies specific for a variety of intestinal bacteria that colonize mucosal surfaces.

Another mechanism of restricting growth of bacteria that penetrate the skin and mucous membranes is competition for iron. Typically, the amount of free iron in tissues and blood available to bacteria is very low, since plasma transferrin binds virtually all iron in the blood. Similarly, hemoglobin in the erythrocytes binds iron. Without free iron, bacterial growth is restricted unless the bacteria synthesize siderophores or receptors for iron containing molecules that compete for transferrin-bound iron. Such siderophores strip iron from transferrin and present it to the bacteria, which enables them to grow. The phagocytic cells of the body patrol the blood and tissues for foreign substances, including bacteria. This task is assumed predominantly by polymorphonuclear neutrophils; however, monocytes, macrophages, and eosinophils also participate.

After phagocytosis, these bacterial cells usually are killed unless their numbers are excessive or they possess virulence factors, that enable them to survive the lysosomal enzymes and acidic pH. In some instances, the bacteria kill the phagocyte or multiply within the macrophage, escaping the hostile extracellular environment. When inflammation occurs, phagocytic cells, along with lymphocytes, play an important role in innate immunity to bacterial infections. During the interaction of bacterial cells with macrophages, T cells, and B cells, specific antibody responses and/or cell-mediated immunity develop to protect against reinfection.

Genetic and Molecular Basis for Virulence

Virulence factors in bacteria may be encoded on chromosomal DNA, bacteriophage DNA, plasmids, or transposons in either plasmids or the bacterial chromosome (Fig. 7-2; Table 7-2). For example, the capacity of the Shigella species to invade cells is a property encoded in part on a 140-mega-dalton plasmid. Similarly, the heat-labile enterotoxin (LTI) of *E coli* is plasmid encoded, whereas the heat-labile toxin (LTII) is encoded on the chromosome. Other virulence factors are acquired by bacteria following infection by a particular bacteriophage, which integrates its genome into the bacterial chromosome by the process of lysogeny. Temperate bacteriophages often serve as the basis of toxin production in pathogenic bacteria. Examples include diphtheria toxin production by Corynebacterium diphtheriae, erythrogenic toxin formation by Streptococcus pyogenes, Shiga-like toxin synthesis by *E coli,* and production of botulinum toxin (types C and D) by *Clostridium botulinum*. Other virulence factors are encoded on the bacterial chromosome (e.g., cholera toxin, *Salmonella enterotoxin,* and *Yersinia* invasion factors).

The transfer of genes for antibiotic resistance among bacteria is a significant medical problem, although none of these properties actually confers increased virulence to the bacterium. Rather, they provide the opportunity for resistant bacteria to proliferate and produce other virulence factors in patients who are being treated with an inappropriate antibiotic.

An intriguing question regarding most bacterial protein toxins is the purpose they serve for the bacteriophage or the bacterium carrying them. Several bacterial toxins are enzymes. For example, cholera toxin, diphtheria toxin, Pseudomonas exotoxin A, and pertussis toxin all are NAD+ glycohydrolases that also act as ADP-ribosyltransferases. The toxic effect of these bacterial enzymes on the host is integral to the pathogenesis of the bacterial infections, but the function of the enzymes in the normal bacterial physiology is not known. Of all the protein toxins synthesized by pathogenic bacteria, there are few instances in which the function of the protein to the bacterium is known. It would be unlikely for the bacterium or infecting bacteriophage to expend the energy necessary to synthesize these relatively high-molecular-weight and complex molecules if they offered it no advantage. Frequently the toxicity of these substances is "unintentional" as far as the bacteria are concerned, considering that the primary goal of the microorganisms is to acquire nutrients and multiply rather than to harm the host.

Host-Mediated Pathogenesis

The pathogenesis of many bacterial infections cannot be separated from the host immune response, for much of the tissue damage is caused by the host response rather than by bacterial factors. Classic examples of host response-mediated pathogenesis are seen in diseases such as Gram-negative bacterial sepsis, tuberculosis, and tuberculoid leprosy. The tissue damage in these infections is caused by toxic factors released from the lymphocytes, macrophages, and polymorphonuclear neutrophils infiltrating the site of infection. Often the host response is so intense that host tissues are destroyed, allowing resistant bacteria to proliferate. In lepromatous leprosy, in contrast, the absence of a cellular response to Mycobacterium leprae allows the bacteria to multiply to such large numbers in the skin that they become tightly packed and replace healthy tissue. The molecular basis for this specific immune anergy is poorly understood.

Intracellular Growth

In general, bacteria that can enter and survive within eukaryotic cells are shielded from humoral antibodies and can be eliminated only by a cellular immune response. However, these bacteria must possess specialized mechanisms to protect them from the harsh effects of the lysosomal enzymes encountered within the cell.

Pathogenic bacteria can be grouped into three categories on the basis of their invasive properties for eukaryotic cells. Although some bacteria (e.g., *Rickettsia, Coxiella,* and *Chlamydia*) grow only inside host cells, others (e.g., *Salmonella, Shigella,* and *Yersinia*) are facultative intracellular pathogens, invading cells when it gives them a selective advantage in the host.

Some bacteria survive the intracellular milieu by producing phospholipases to dissolve the phagocytic vesicle surrounding them. This appears to be the case for R rickettsii, which destroys the phagosomal membrane with which the lysosomes fuse. Legionella pneumophila, which prefers the intracellular environment of macrophages for growth, appears to induce its own uptake and blocks lysosomal fusion by undefined mechanisms.

Other bacteria have evolved to the point that they prefer the low-pH environment within the lysosomal granules, as may be the case for *Coxiella burnetii,* a highly resistant member of the rickettsial group. Salmonella and Mycobacterium species also appear to be very resistant to intracellular killing by phagocytic cells, but their mechanisms of resistance are not yet fully understood. Certainly, the capacity of bacteria to survive and multiply within host cells has great impact on the pathogenesis of the respective infections.

Most bacterial pathogens do not invade cells, proliferating instead in the extracellular environment enriched by body fluids. Some of these bacteria (e.g., V cholerae and Bordetella pertussis) do not even penetrate body tissues, but, rather, adhere to epithelial surfaces and cause disease by secreting potent protein toxins. Although bacteria such as *E coli* and *P aeruginosa* are termed noninvasive, they frequently spread rapidly to various tissues once they gain access to the body. All bacteria could at some point be considered intracellular once they become ingested by polymorphonuclear neutrophils and macrophages, but these organisms are not renowned for their capacity to survive the intracellular environment or to induce their own uptake by most host cells.

Specific Virulence Factors

The virulence factors of bacteria can be divided into a number of functional types. These are discussed in the following sections:

Adherence and Colonization Factors

To cause infection, many bacteria must first adhere to a mucosal surface. For example, the alimentary tract mucosa is continually cleansed by the release of mucus from goblet cells and by the peristaltic flow of the gut contents over the epithelium. Similarly, ciliated cells in the respiratory tract sweep mucus and bacteria upward. In addition, the turnover of epithelial cells at these surfaces is fairly rapid. The intestinal epithelial cell monolayer is continually replenished, and the cells are pushed from the crypts to the villar tips in about 48 hours.

To establish an infection at such a site, a bacterium must adhere to the epithelium and multiply before the mucus and extruded epithelial cells are swept away. To accomplish this, bacteria have evolved attachment mechanisms, such as pili (fimbriae), that recognize and attach the bacteria to cells. Colonization factors (as they are often called) are produced by numerous bacterial pathogens and constitute an important part of the pathogenic mechanism of these bacteria. Some examples of piliated, adherent bacterial pathogens are *V cholerae, E coli, Salmonella* spp, *N gonorrheae, N meningitidis,* and *Streptococcus pyogenes.*

Invasion Factors

Mechanisms that enable a bacterium to invade eukaryotic cells facilitate entry at mucosal surfaces. Some of these invasive bacteria (such as Rickettsia and Chlamydia species) are obligate intracellular pathogens, but most are facultative intracellular pathogens. The specific bacterial surface factors that mediate invasion are not known in most instances, and often, multiple gene products are involved. Some *Shigella* invasion factors are encoded on a 140 megadalton plasmid, which, when conjugated into *E coli,* gives these noninvasive bacteria the capacity to invade cells. Other invasion genes have also recently been identified in *Salmonella* and *Yersinia* pseudotuberculosis. The mechanisms of invasion of *Rickettsia,* and *Chlamydia* species are not well known.

Capsules and Other Surface Components

Bacteria have evolved numerous structural and metabolic virulence factors that enhance their survival rate in the host. Capsule formation has long been recognized as a protective mechanism for bacteria. Encapsulated strains of many bacteria (e.g., pneumococci) are more virulent and more resistant to phagocytosis and intracellular killing than are nonencapsulated strains.

Organisms that cause bacteremia (e.g., *Pseudomonas*) are less sensitive than many other bacteria to killing by fresh human serum containing complement components, and consequently are called serum resistant. Serum resistance may be related to the amount and composition of capsular antigens as well as to the structure of the lipopolysaccharide. The relationship between surface structure and virulence is important also in *Borrelia* infections. As the bacteria encounter an increasing specific immune response from the host, the bacterial surface antigens are altered by mutation, and the progeny, which are no longer recognized by the immune response, express renewed virulence. *Salmonella typhi* and some of the paratyphoid organisms carry a surface antigen, the *Vi antigen*, thought to enhance virulence. This antigen is composed of a polymer of galactosamine and uronic acid in 1,4-linkage. Its role in virulence has not been defined, but antibody to it is protective. Some bacteria and parasites have the ability to survive and multiply inside phagocytic cells. A classic example is Mycobacterium tuberculosis, whose survival seems to depend on the structure and composition of its cell surface. The parasite *Toxoplasma gondii* has the remarkable ability to block the fusion of lysosomes with the phagocytic vacuole. The hydrolytic enzymes contained in the lysosomes are unable, therefore, to contribute to the destruction of the parasite. The mechanism(s) by which bacteria such as *Legionella pneumophila, Brucella abortus,* and *Listeria* monocytogenes remain unharmed inside phagocytes are not understood.

Endotoxins

Endotoxin is comprised of toxic lipopolysaccharide components of the outer membrane of Gram-negative bacteria. Endotoxin exerts profound biologic effects on the host and may be lethal. Because it is omnipresent in the environment, endotoxin must be removed from all medical supplies destined for injection or use during surgical procedures. The term endotoxin was coined in 1893 by Pfeiffer to distinguish the class of toxic substances released after lysis of bacteria from the toxic substances (exotoxins) secreted by bacteria. Few, if any, other microbial products have been as extensively studied as bacterial endotoxins. Perhaps it is appropriate that a molecule with such important biologic effects on the host, and one produced by so many bacterial pathogens, should be the subject of intense investigation.

Structure of Endotoxin

The complex is secured to the outer membrane by ionic and hydrophobic forces, and its strong negative charge is neutralized by Ca^{2+} and Mg^{2+} ions.

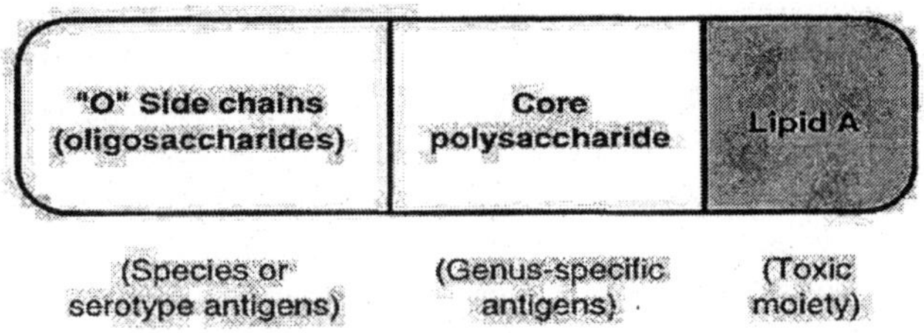

FIGURE: Basic structure of endotoxin (lipopolysaccharide) from Gram-negative bacteria.

The structure of endotoxin molecules from *Salmonella* spp and *E coli* is known in detail. Enough data on endotoxin from other Gram-negative organisms have been gathered to reveal a common pattern with genus and species diversity. Although all endotoxin molecules are similar in chemical structure and biologic activity, some diversity has evolved. Purified endotoxin appears as large aggregates.

The molecular complex can be divided into three regions: (1) the O-specific chains, which consist of a variety of repeating oligosaccharide residues, (2) the core polysaccharide that forms the backbone of the macromolecule, and (3) lipid A, composed usually of a glucosamine disaccharide with attached long-chain fatty acids and phosphate.

The polysaccharide portions are responsible for antigenic diversity, whereas the lipid A moiety confers toxicity. Dissociation of the complex has revealed that the polysaccharide is important in solubilizing the toxic lipid A component, and in the laboratory it can be replaced by carrier proteins (e.g., bovine serum albumin).

Members of the family Enterobacteriaceae exhibit O-specific chains of various lengths, whereas *N gonorrhoeae, N meningitidis,* and *B pertussis* contain only core polysaccharide and lipid A. Some investigators working on the latter forms of endotoxin prefer to call them lipooligosaccharides to emphasize the chemical difference from the endotoxin of the enteric bacilli. Nevertheless, the biologic activities of all endotoxin preparations are essentially the same, with some being more potent than others.

Biologic Activity of Endotoxin

The biologic effects of endotoxin have been extensively studied. Purified lipid A (conjugated to bovine serum albumin) and endotoxin elicit the same biologic responses. The more pertinent toxic effects include pyrogenicity, leukopenia followed by leukocytosis, complement activation, depression in blood pressure, mitogenicity, induction of prostaglandin synthesis, and hypothermia. These events can culminate in sepsis and lethal shock.These include:

1. mitogenic effects on B lymphocytes that increase resistance to viral and bacterial infections
2. induction of gamma interferon production by T lymphocytes, which may enhance the antiviral state, promote rejection of tumor cells, and activate macrophages and natural killer cells
3. activation of the complement cascade with the formation of C3a and C5a
4. induction of the formation of interleukin-1 by macrophages and interleukin-2 and other mediators by T lymphocytes.

Current research focuses on exploiting some of the potential beneficial effects of "nontoxic" endotoxin derivatives and holds promise for development of future treatment regimens for stimulating the immune response. For example, the toxicity of endotoxin is largely attributed to lipid A, attached to a polysaccharide carrier. The toxicity of lipid A is markedly reduced after hydrolysis of a phosphate group or deacylation of one or more fatty acids from the lipid A molecule. Clinical trials are in progress to test a monophosphoryl lipid A for its potential of inducing low dose tolerance to endotoxin. Tolerance to endotoxin can be achieved by pretreatment of an animal with low doses of endotoxin or a detoxified lipid A derivative before challenge with high doses of endotoxin. Experimental studies have demonstrated that induction of tolerance to endotoxin reduces the dangerous effects of endotoxin. It is hoped that these relatively nontoxic lipid A derivatives may be useful in reducing the severity of bacterial sepsis in which bacterial endotoxin produces a life-threatening clinical course. Endotoxin, which largely accumulates in the liver following injection of a sublethal dose by the intravenous route, can be devastating because of its ability to affect a variety of cell and host proteins.

Kupffer cells, granulocytes, macrophages, platelets, and lymphocytes all have a cell receptor on their surface called CD14, which binds endotoxin. ndotoxin binding to the CD14 receptor on macrophages is enhanced by interaction with a host protein made in the liver (i.e., LPS-binding protein). The extent of involvement of each cell type probably depends on the level of endotoxin exposure.

The effects of endotoxin on such a wide variety of host cells result in a complex array of host responses that can culminate in the serious condition gram-negative sepsis, which often leads to shock and death. The effects of endotoxin on host cells are known to stimulate prostaglandin synthesis and to activate the kallikrein system, the kinin system, the complement cascade via the alternative pathway, the clotting system, and the fibrinolytic pathways.

When these normal host systems are activated and operate out of control, it is not surprising that endotoxin can be lethal. Although it is difficult to comprehend the mechanisms of all the cell responses and the myriad sequelae of the cell mediators released rather indiscriminately in the host following exposure to endotoxin, it does seem clear that the host cellular response to endotoxin, rather than a direct toxic effect of endotoxin, plays the major role in causing tissue damage.

DETECTION OF ENDOTOXIN IN MEDICAL SOLUTIONS

Endotoxin is omnipresent in the environment. It is found in most deionized-water lines in hospitals and laboratories, for example, and affects virtually every biologic assay system ever examined. It tends to be a scapegoat for all biologic problems encountered in the laboratory, and, many times, this reputation is deserved. Because of its pyrogenic and destructive properties, extreme care must be taken to avoid exposing patients to medical solutions containing endotoxin. Even though all supplies should be sterile, solutions for intravenous administration can become contaminated with endotoxin-containing bacteria after sterilization as a result of improper handling.

Furthermore, water used in the preparation of such solutions must be filtered through ion exchange resins to remove endotoxin, because it is not removed by either autoclave sterilization or filtration through bacterial membrane filters. If endotoxin-containing solutions were used in such medical procedures as renal dialysis, heart bypass machines, blood transfusions, or surgical lavage, the patient would suffer immediate fever accompanied by a rapid and possibly lethal alterations in blood pressure.

Solutions for human or veterinary use are prepared under carefully controlled conditions to ensure sterility and to remove endotoxin. Representative samples of every manufacturing batch are checked for endotoxin by one of two procedures: the Limulus lysate test or the rabbit pyrogenicity test. The rabbit pyrogenicity test is based on the exquisite sensitivity of rabbits to the pyrogenic effects of endotoxin. A sample of the solution to be tested usually is injected intravenously into the ear veins of adult rabbits while the rectal temperature of the animal is monitored. Careful monitoring of the temperature responses provides a sensitive and reliable indicator of the presence of endotoxin and, importantly, one measure of the safety of the solution for use in patients.

The Limulus lysate test is more common and less expensive. This test, which is based on the ability of endotoxin to induce gelation of lysates of amebocyte cells from the horseshoe crab Limulus polyphemus, is simple, fast, and sensitive (about 1 ng/ml). It is so sensitive, however, that trace quantities of endotoxin in regular deionized water often obscure the results.

It can be used for rapid detection of certain Gram-negative infections (e.g., of cerebrospinal fluid); however, blood contains inhibitors that prevent gelation. Test kits are commercially available. The amebocyte is the sole phagocytic immune cell of the horseshoe crab, and the gelation reaction is believed to be involved in sequestering invading Gram-negative bacteria.

Exotoxins

Exotoxins, unlike the lipopolysaccharide endotoxin, are protein toxins released from viable bacteria. They form a class of poisons that is among the most potent, per unit weight, of all toxic substances. Most of the higher molecular-sized exotoxin proteins are heat labile; however, numerous low molecular-sized exotoxins are heat-stable peptides. Unlike endotoxin, which is a structural component of all Gram-negative cells, exotoxins are produced by some members of both Gram-positive and Gram-negative genera.

The functions of these exotoxins for the bacteria are usually unknown, and the genes for most can be deleted with no noticeable effect on bacterial growth. In contrast to the extensive systemic and immune-system effects of endotoxin on the host, the site of action of most exotoxins is more localized and is confined to particular cell types or cell receptors. Tetanus toxin, for example, affects only internuncial neurons. In general, exotoxins are excellent antigens that elicit specific antibodies called antitoxins. Not all antibodies to exotoxins are protective, but some react with important binding sites or enzymatic sites on the exotoxin, resulting in complete inhibition of the toxic activity (i.e., neutralization).

Exotoxins can be grouped into several categories (e.g., neurotoxins, cytotoxins, and enterotoxins) based on their biologic effect on host cells. Neurotoxins are best exemplified by the toxins produced by *Clostridium* spp, for example, the botulinum toxin formed by *C botulinum*. This potent neurotoxin acts on motor neurons by preventing the release of acetylcholine at the myoneural junctions, thereby preventing muscle excitation and producing flaccid paralysis.

The cytotoxins constitute a larger, more heterogeneous grouping with a wide array of host cell specificities and toxic manifestations. One cytotoxin is diphtheria toxin, which is produced by *Corynebacterium diphtheriae*. This cytotoxin inhibits protein synthesis in many cell types by catalyzing the ADP-ribosylation of elongation factor II, which blocks elongation of the growing peptide chain.

Enterotoxins stimulate hypersecretion of water and electrolytes from the intestinal epithelium and thus produce watery diarrhea. Some enterotoxins are cytotoxic (e.g., shiga-like enterotoxin from *E coli*), while others perturb eukaryotic cell functions and are cytotonic (e.g., cholera toxin).

Enterotoxins also can disturb normal smooth muscle contraction, causing abdominal cramping and decrease transit time for water absorption in the intestine. Enterotoxigenic *E coli* and *V cholerae* produce diarrhea after attaching to the intestinal mucosa, where they elaborate enterotoxins. Neither pathogen invades the body in substantial numbers, except in the case of *E coli* species that have acquired an invasion plasmid.

Importantly, cholera toxin and *E coli* heat-labile enterotoxins I and II cause ADP-ribosylation of cell proteins in a manner similar to diphtheria toxin, except that the primary target is the regulatory protein (G_s) of adenylate cyclase, resulting in increased levels of cyclic 3',5'-adenosine monophosphate (cAMP).

In contrast, the organisms responsible for shigellosis (*Shigella dysenteriae, S boydii, S flexneri,* and *S sonnei*) penetrate the mucosal surface of the colon and terminal ileum to proliferate and cause ulcerations that bleed into the intestinal lumen. Despite causing extensive ulceration of the mucosa, the pathogens rarely enter the bloodstream. The Shiga enterotoxin produced by *Shigella* species and the Shiga-like enterotoxin elaborated by many isolates of *E coli* inhibit protein synthesis in eukaryotic cells.

It is not clear how this cytotoxic enterotoxin causes hypersecretion of water and electrolytes from the intestinal epithelium. These enterotoxins differ from those secreted by *V cholerae* and *E coli* in that the Shiga toxins are cytotoxic and lethal, whereas the cholera toxin-like enterotoxins are not. The latter enterotoxins cause no structural damage to cells, and are described as cytotonic. The ensuing inflammatory response to the invading bacteria and/or their toxins appears to activate neurologic control mechanisms (e.g., prostaglandins, serotonin) that normally regulate water and electrolyte transport.

Siderophores

Both animals and bacteria require iron for metabolism and growth, and the control of this limited resource is often used as a tactic in the conflict between pathogen and host. Animals have evolved mechanisms of "withholding" iron from tissue fluids in an attempt to limit the growth of invading bacteria.

Although blood is a rich source of iron, this iron is not readily available to bacteria since it is not free in solution. Most of the iron in blood is bound either to hemoglobin in erythrocytes or to transferrin in plasma. Similarly, the iron in milk and other secretions (e.g., tears, saliva, bronchial mucus, bile, and gastrointestinal fluid) is bound to lactoferrin. Some bacteria express receptors for eukoyotic iron-binding proteins (e.g., transferrin-binding outer membrane proteins on the surface of *Neisserira* spp). Via these specialized receptors iron acquisition is facilitated, providing the esssential element for bacterial growth.

Other bacteria have evolved elaborate mechanisms to extract the iron from host proteins. Siderophores are substances produced by many bacteria (and some plants) to capture iron from the host. The absence of iron triggers transcription of the genes coding for the enzymes that synthesize siderophores, as well as for a set of surface protein receptors that recognize siderophores carrying bound iron. The binding constants of the siderophores for iron are so high that even iron bound to transferrin and lactoferrin is confiscated and taken up by the bacterial cells.

An example of a bacterial siderophore is enterochelin, which is produced by *Escherichia* and *Salmonella* species. Classic experiments have demonstrated that Salmonella mutants that have lost the capacity to synthesize enterochelin lose virulence in an assay of lethality in mice. Injection of purified enterochelin along with the Salmonella mutants restores virulence to the bacteria. Therefore, siderophore production by many pathogenic bacteria is considered an important virulence mechanism.

Epilogue

Many factors determine the outcome of the bacterium-host relationship. The host must live in an environment filled with a diverse population of microorganisms. Because of the magnitude of the infectious-disease problem, we strive to understand the natural immune mechanisms of the host so that future improvements in resistance to bacterial infections may be possible. Similarly, massive research efforts are being expended to identify and characterize the virulence factors of pathogenic bacteria and hence allow us to interrupt the pathogenic mechanisms of virulent bacteria. The availability of an array of antibiotics and vaccines has provided the medical profession with powerful tools to control or cure many infections. Unfortunately, these drugs and vaccines have eliminated no bacterial disease from the human or animal populations, and bacterial infections and drug resistance remain a serious medical problem.

6

Specific Acquired Immunity

INTRODUCTION

Basis of Acquired Resistance

Acquired resistance is mediated by antigen-specific immune mechanisms. This specificity may be acquired following a disease, by asymptomatic carriage of the pathogen, by harboring an organism with a similar structure (crossreacting,) or by vaccination.

Specific acquired immunity against infectious diseases may be mediated by antibodies and/or T lymphocytes. Immunity mediated by these two factors may be manifested by a direct effect upon a pathogen, such as (1) antibody-initiated complement-dependent bacteriolysis, (2) opsonophagocytosis and killing, as occurs for some bacteria, (3) neutralization of viruses so that these organisms cannot enter cells, or (4) by T lymphocytes which will kill a cell parasitized by a microorganism.

Primary vs Opportunistic Pathogens

Among the almost infinite varieties of microorganisms, relatively few cause a disease in an otherwise normal or healthy individual. These highly virulent microorganisms are conveniently classified as primary pathogens. Opportunistic infectious disease may be caused by organisms that are ordinarily in contact with the host, such as bacteria or fungi in the colon or in the upper respiratory tract; following an injury, whether mechanical (such as a open fracture); or by a disease with immunosuppressive activity (such as measles or malaria, or one induced by cytotoxic chemotherapy). Organisms, which cause an infectious disease in a host with depressed resistance, are classified as opportunistic pathogens. Primary pathogens may also cause more explosive disease in a host with depressed resistance. Our knowledge of the protective antigens and specific acquired host immune factors is more complete for primary pathogens.

Most primary pathogens are inhabitants of, and pathogens for, humans only. Opportunistic pathogens, in contrast, may cause disease in many species of mammals. Some exceptions are tetanus, anthrax and rabies which may inhabit and cause disease in many animal species including humans.

Protective Antigens

Microorganisms adapt to cause disease by many mechanisms. Many bacteria, for example, produce macromolecules that cause (1) inflammation, (2) adherence to human tissues, or (3) are toxins that chemically alter host metabolism. But, in general, specific acquired immunity to human pathogens is directed to one (protective) antigen.

Protein Antigens

Proteins, in most instances, are defined as T cell dependent antigens by two properties. First, proteins are hydrolyzed to peptides by intracellular proteases. Second, proteins may have multiple but unique specificities (epitopes). These two properties permit recognition by receptors (membrane immunoglobulin) of antibody-producing cells (plasma cells) and internalization and proteolysis by enzymes. Activation requires that the peptide fragment of the protein antigen interact with the plasma cell histocompatibility antigen to form a complex that attracts and activates T cells. The activated antigen-specific T cells release cytokines that cause plasma cells to divide and to increase their secretion of specific antibody.

The same mechanism serves to activate T cells by contact with host cells parasitized by intracellular microorganisms. Proteins secreted by intracellular organisms interact with histocompatibility antigens of the host cell (T cell epitopes) and provide a specific site for T cells that kill the parasitized host cell (cytotoxic T lymphocytes).

Some protective epitopes, especially those of viruses, are expressed only on the intact organism (conformation epitope). An example is the neutralizing epitope of polioviruses (D antigen) that requires the intact capsid to elicit neutralizing antibodies.

The protective protein antigens are highly specific and are unique to each pathogen. Acquisition of antibodies to protein protective antigens either follows infection with the pathogen or vaccination.

Polysaccharide Antigens

Surface polysaccharides of pathogens may serve as protective antigens. These polysaccharides may be capsular polysaccharides, present on either Gram-negative or Gram-positive organisms or the outermost domain of the lipopolysaccharide of Gram-negative organisms. Polysaccharides have simple structures composed of identical repeating units so that each molecule will have relatively fewer epitopes than a protein. But in contrast to a protein, a polysaccharide is multivalent for each epitope. This multivalency explains why a polysaccharide can crosslink the receptors of plasma cells, resulting in activation and multiplication of plasma cells, both of which increase secretion of antibodies.

The protective epitopes of polysaccharides, in contrast to proteins, are widely shared in nature, and natural immunity: antibody synthesis in the absence of the homologous organism, occurs in almost every individual during development. Similarly, disease and often asymptomatic carriage will also stimulate serum polysaccharide antibodies. Lastly, because they do not interact with T cells, polysaccharides are designated as T cell independent.

Each bacterial species, such as pneumococci, may have many capsular polysaccharides but only a fraction of these will be associated with a disease. For example, there are now 89 reported types but most systemic pneumococcal infections are caused by about 23 types. In infants and children, most pneumococcal infections are caused by only 8 types. Similarly, of the six types of Haemophilus influenzae, almost all systemic infections, especially meningitis, are caused by type b.

It is important to understand that immunity may be directed towards the intact pathogens such as bacteria, viruses, protozoa, or fungi, or to individual extracellular antigens such as toxins (antitoxin).

IMMUNE MECHANISMS

Antibody and Secondary Biologic Activities

Serum antibodies are the signal and specific component of a complex inactivation system. Serum IgM and IgG antibodies exert their protective effect directly upon bacteria whose surface polysaccharides or proteins are protective antigens. This signal is generated by the configuration of the non-antibody combining site region of the heavy polypeptide chain after binding of antibody with an epitope. For some pathogens such as meningococci (Gram-negative), the antigen-antibody configuration will activate the serum complement protein cascade with deposition of a C8,9 peptide probe that drills itself through the outer membrane. The resultant lesion causes release of intracellular components and lysis of the meningococci. In addition to bacteriolysis, other Gram-negatives, such as Haemophilus influenzae type b, also may be inactivated by serum antibody, affixed to the polysaccharide of this pathogen, that attracts and activates serum complement proteins to form C3 and C5 complexes. The latter complexes attract and activate phagocytic cells that engulf and digest the pathogen (opsonophagocytosis). This antibody-initiated, complement-dependent phagocytosis and killing are required for Gram-positive bacteria whose cell wall is not susceptible to lysis by complement.

Serum antibodies may confer immunity by binding directly to viral pathogens. The effect of this simple interaction is neutralization or inactivation of the virus. The exact protective mechanism of viral-specific antibodies is not known and may be unique for each pathogen. In many cases, antibody binding renders the viral pathogen incapable of infecting a host cell by preventing penetration of cells. As an example, antibodies to the fusion (F) protein of measles prevent the integration of virus with the cell membrane of the host. Experimental proof for this direct antiviral effect is that monovalent fragments of IgG, unable to activate complement, exert similar neutralization activity as the intact antibody. Some larger viral pathogens, coated with specific antibody, may be phagocytized and digested.

Antibodies that neutralize bacterial toxins (antitoxins), such as tetanus, diphtheria and pertussis toxins, are protective and therapeutic. The protective effects of antitoxins are varied. Antitoxin does not exert antibacterial action upon Clostridium tetani. Rather, antitoxin inactivates the functions of tetanus toxin that facilitates its migration up the neural sheath to the synapse, and antibodies to the enzymatic region inhibit its alteration of the synapse. The neutralizing activities of diphtheria and pertussis antitoxins, in contrast, exert secondary antibacterial actions upon their respective pathogens. Both toxins serve to condition the respiratory epithelium to permit colonization by Corynebacterium diphtheriae tox+ and Bordetella pertussis: the former by its cytotoxicity and the latter by its inactivation of the function of phagocytic cells. Antitoxins block these actions and facilitate the function of phagocytic cells.

Cell-mediated Immunity

Antigen-specific activation of T cells has been described (vide supra). The targets for activated T cells are parasitized host cells. The secondary or inactivation mechanisms invoked by activated T cell phagocytic host cell complexes are not clearly understood. One important mechanism is the release of nitrous oxide that results in killing of the host cell and of the pathogen. Cell-mediated immunity is largely, if not exclusively, a curative mechanism.

Preventive Immunity

Immune resistance to an infectious disease requires a critical level of either antigen-specific antibodies and/or T cells when the host encounters the pathogen. Prevention of an infection requires immune mechanisms to kill or inactivate the inoculum of the pathogen. This immunity may be expressed as a protective level of antibodies so that resistance to specific infections may be reliably predicted by a serologic assay such as the level of neutralizing antibodies to measles, mumps or Groups A, B, Y and W135 meningococci. These assays can predict resistance to a disease for individuals or can be used to assess the immune status of communities. Quantitation of antigen-specific T cells to predict immunity on a clinical basis is, as yet, an investigative tool.

Disease-acquired Immunity

Convalescence from most infectious diseases confers immunity. This immunity, in most instances, may be transferred for a limited period to non-immune individuals by injection of IgG as FDA-licensed immunoglobulin or to the newborn by passively acquired maternal serum IgG. There is also evidence that secretory IgA acquired by breast feeding confers immunity to newborns. These findings indicate that critical levels of antibodies are sufficient to prevent infectious diseases. Their preventive action is best explained by antibodies killing or inactivating the inoculum of the pathogen on epithelial surfaces. Herd immunity follows vaccination with Haemophilus type b conjugates, diphtheria toxoid and measles virus vaccines. The resulting immunity causes a decreased transmission of the pathogen. Since there is no animal vector for these pathogens, the incidence of the disease in the entire community is far below that percentage of the population that has been vaccinated (herd immunity). Antibody-mediated inactivation of the inoculum may also occur in the blood stream in the case of pathogens inoculated directly into the tissues or blood stream, such as hepatitis B or malaria. There is yet no evidence in humans that antigen-specific T lymphocytes can prevent infectious diseases.

Natural Immunity

Acquisition of serum antibodies to surface polysaccharides of pathogens is age-related and often occurs without the individual encountering the homologous organism. An example is group A meningococci, the cause of epidemic meningitis. Despite the absence of this pathogen in the United States for about 50 years, either as a cause of meningitis or in asymptomatic carriers, most adults have antibodies to this capsular polysaccharide. Antigenic stimuli for Group A meningococcal antibodies are likely due to exposure to several Gram-positive and Gram-negative bacterial species in human stools. Another example is Shigella dysenteriae type 1, the cause of epidemic dysentery. Adults in Sweden and the United States have antibodies to the LPS of S dysenteriae type 1, despite the virtual absence of this pathogen in these countries during the past 50 years. The stimulus for these natural and protective antibodies is probably cross-reacting polysaccharides of the enteric and respiratory tract floras.

Natural serum antibodies to surface polysaccharides confer specific protection to adults and are transmitted to newborns. The highest attack rate and mortality occur during childhood when these maternally-acquired anti-polysaccharide antibodies have waned and adult levels have not been reached. Acquisition of natural antibodies is not uniform and many adults remain non-immune. Vaccination with polysaccharide-based vaccines increases the percentage of adults with protective levels to almost 100%. These principles are elegantly illustrated by the development of groups A and C meningococcal polysaccharide vaccines during the 1960's when outbreaks of meningitis caused by this pathogen occurred in armed forces recruits during the Vietnam conflict. Introduction of these polysaccharide vaccines rapidly eliminated these outbreaks.

Vaccination-induced Active and Passive Immunity

Vaccines are heterogeneous according to (1) the nature of the immunizing substance, whether they are inert or living, and (2) by the method of their administration. Vaccines include living attenuated strains of viruses (poliovirus) or bacteria (BCG), inactivated viruses (yellow fever) and bacteria (anthrax), purified polysaccharides (pneumococcal 23 valent) or polysaccharide-protein conjugates (Haemophilus type b conjugate). Inert and injected vaccines, such as tetanus toxoid, elicit mostly serum antibodies. Living vaccines, such as attenuated strains of viruses (poliovirus), elicit secretory antibodies and sensitized T cells.

To date, the FDA regulates vaccines and seroepidemiologic studies to assess the immune status of populations by measurement of biologically active antibodies only. Thus, the status of diphtheria immunity is evaluated by the percentage of the population with protective levels of neutralizing antibodies to diphtheria toxin (antitoxin). Similarly, the immune status to measles is evaluated by measurement of the percentage of the population with protective levels of neutralizing antibodies. For some vaccines, such as BCG, there is as yet no measure of immunity to assess their effectiveness.

Some investigational vaccines utilize recombinant DNA technology to mobilize genes governing the synthesis of protective antigens. These specific genes may be inserted into avirulent vectors. Administration of vectors is designed to stimulate the comprehensive immunity that follows disease with the pathogen itself. Naked DNA may be incorporated into plasmids that infect somatic cells and continually induce synthesis of protective antigens that stimulate antibodies and activated T cells.

Curative Immunity

Infection with most pathogens does not result in death of the host and the offending organism is ultimately cleared after the symptoms of the disease have waned. The basis of the curative process of patients is not well understood, but it is likely mediated by expansion of both specific immune and effector mechanisms. Quantitative increases of specific antibodies and activated T cells is accompanied by increases in serum complement levels and phagocytic cells. Also, infection increases the levels of cytokines and serum proteins known as acute phase reactants, such as C-reactive protein, alpha 1 trypsin inhibitor, and transferrin, that serve as scavengers or inhibitors of bacterial debris. Cure of infectious diseases is most likely the prolonged interaction of maximal levels of host specific and non-specific (effector) mechanisms with the pathogen.

7

Epidemiology

INTRODUCTION

This chapter reviews the general concepts of epidemiology, which is the study of the determinants, occurrence, distribution, and control of health and disease in a defined population. Epidemiology is a descriptive science and includes the determination of rates, that is, the quantification of disease occurrence within a specific population. The most commonly studied rate is the attack rate: the number of cases of the disease divided by the population among whom the cases have occurred. Epidemiology can accurately describe a disease and many factors concerning its occurrence before its cause is identified. For example, Snow described many aspects of the epidemiology of cholera in the late 1840s, fully 30 years before Koch described the bacillus and Semmelweis described puerperal fever in detail in 1861 and recommended appropriate control and prevention measures a number of years before the streptococcal agent was fully described. One goal of epidemiologic studies is to define the parameters of a disease, including risk factors, in order to develop the most effective measures for control. This chapter includes a discussion of the chain of infection, the three main epidemiologic methods, and how to investigate an epidemic.

Proper interpretation of disease-specific epidemiologic data requires information concerning past as well as present occurrence of the disease. An increase in the number of reported cases of a disease that is normal and expected, representing a seasonal pattern of change in host susceptibility, does not constitute an epidemic. Therefore, the regular collection, collation, analysis, and reporting of data concerning the occurrence of a disease is important to properly interpret short-term changes in occurrence.

A sensitive and specific surveillance program is important for the proper interpretation of disease occurrence data. Almost every country has a national disease surveillance program that regularly collects data on selected diseases. The quality of these programs varies, but, generally, useful data are collected that are important in developing control and prevention measures.

There is an international agreement that the occurrence of three diseasescholera, plague, and yellow feverwill be reported to the World Health Organization in Geneva, Switzerland. In the United States, the Centers for Disease Control and Prevention (CDC), U.S. Public Health Service, and the state health officers of all 50 states have agreed to report the occurrence of 51 diseases weekly and of another 10 diseases annually from the states to the CDC. Many states have regulations or laws that mandate reporting of these diseases and often of other diseases of specific interest to the state health department.

The methods of case reporting vary within each state. Passive reporting is one of the main methods. In such a case, physicians or personnel in clinics or hospitals report occurrences of relevant diseases by telephone, postcard, or a reporting form, usually at weekly intervals. In some instances, the report may be initiated by the public health or clinical laboratory where the etiologic agent is identified. Some diseases, such as human rabies, must be reported by telephone as soon as diagnosed. In an active surveillance program, the health authority regularly initiates the request for reporting. The local health department may call all or some health care providers at regular intervals to inquire about the occurrence of a disease or diseases. The active system may be used during an epidemic or if accurate data concerning all cases of a disease are desired.

The health care provider usually makes the initial passive report to a local authority, such as a city or county health department. This unit collates its data and sends a report to the next highest health department level, usually the state health department.

The number of cases of each reportable disease are presented weekly, via computer linkage, by the state health department to the CDC. Data are analyzed at each level to develop needed information to assist public health authorities in disease control and prevention. For some diseases, such as hepatitis, the CDC requests preparation of a separate case reporting form containing more specific details.

In addition, the CDC prepares and distributes routine reports summarizing and interpreting the analyses and providing information on epidemics and other appropriate public health matters. Most states and some county health departments also prepare and distribute their own surveillance reports. The CDC publishes *Morbidity and Mortality Weekly Report,* which is available for a small fee from the Massachusetts Medical Society. The CDC also prepares more detailed surveillance reports for specific diseases, as well as an annual summary report, all of which can also be obtained through the Massachusetts Medical Society.

Infection is the replication of organisms in the tissue of a host; when defined in terms of infection, disease is overt clinical manifestation. In an inapparent (subclinical) infection, an immune response can occur without overt clinical disease. A carrier (colonized individual) is a person in whom organisms are present and may be multiplying, but who shows no clinical response to their presence.

The carrier state may be permanent, with the organism always present; intermittent, with the organism present for various periods; or temporary, with carriage for only a brief period. Dissemination is the movement of an infectious agent from a source directly into the environment; when infection results from dissemination, the source, if an individual, is rcferred to as a dangerous disseminator.

Infectiousness is the transmission of organisms from a source, or reservoir (see below), to a susceptible individual. A human may be infective during the preclinical, clinical, postclinical, or recovery phase of an illness. The incubation period is the interval in the preclinical period between the time at which the causative agent first infects the host and the onset of clinical symptoms; during this time the agent is replicating.

Transmission is most likely during the incubation period for some diseases such as measles; in other diseases such as shigellosis, transmission occurs during the clinical period. The individual may be infective during the convalescent phase, as in diphtheria, or may become an asymptomatic carrier and remain infective for a prolonged period, as do approximately 5% of persons with typhoid fever.

The spectrum of occurrence of disease in a defined population includes sporadic (occasional occurrence); endemic (regular, continuing occurrence); epidemic (significantly increased occurrence); and pandemic (epidemic occurrence in multiple countries).

Chain of Infection

The chain of infection includes the three factors that lead to infection: the etiologic agent, the method of transmission, and the host. These links should be characterized before control and prevention measures are proposed. Environmental factors that may influence disease occurrence must be evaluated.

Etiologic Agent

The etiologic agent may be any microorganism that can cause infection. The pathogenicity of an agent is its ability to cause disease; pathogenicity is further characterized by describing the organism's virulence and invasiveness. Virulence refers to the severity of infection, which can be expressed by describing the morbidity (incidence of disease) and mortality (death rate) of the infection. An example of a highly virulent organism is *Yersinia pestis,* the agent of plague, which almost always causes severe disease in the susceptible host.

The invasiveness of an organism refers to its ability to invade tissue. *Vibrio cholerae* organisms are noninvasive, causing symptoms by releasing into the intestinal canal an exotoxin that acts on the tissues. In contrast, *Shigella* organisms in the intestinal canal are invasive and migrate into the tissue.

No microorganism is assuredly avirulent. An organism may have very low virulence, but if the host is highly susceptible, as when therapeutically immunosuppressed, infection with that organism may cause disease. For example, the poliomyelitis virus used in oral polio vaccine is highly attenuated and thus has low virulence, but in some highly susceptible individuals it may cause paralytic disease. Other factors should be considered in describing the agent. The infecting dose (the number of organisms necessary to cause disease) varies according to the organism, method of transmission, site of entrance of the organism into the host, host defenses, and host species. Another agent factor is specificity; some agents (for example, *Salmonella typhimurium)* can infect a broad range of hosts; others have a narrow range of hosts. *S typhi,* for example, infects only humans. Other agent factors include antigenic composition, which can vary within a species (as in influenza virus or *Streptococcus* species); antibiotic sensitivity; resistance transfer plasmids; and enzyme production. The reservoir of an organism is the site where it resides, metabolizes, and multiplies. The source of the organism is the site from which it is transmitted to a susceptible host, either directly or indirectly through an intermediary object. The reservoir and source can be different; for example, the reservoir for *S typhi* could be the gallbladder of an infected individual, but the source for transmission might be food contaminated by the carrier. The reservoir and source can also be the same, as in an individual who is a permanent nasal carrier of *S aureus* and who disseminates organisms from this site. The distinction can be important when considering where to apply control measures.

Method of Transmission

The method of transmission is the means by which the agent goes from the source to the host. The four major methods of transmission are by contact, by common vehicle, by air or via a vector.

In contact transmission the agent is spread directly, indirectly, or by airborne droplets. Direct contact transmission takes place when organisms are transmitted directly from the source to the susceptible host without involving an intermediate object; this is also referred to as person-to-person transmission. An example is the transmission of hepatitis A virus from one individual to another by hand contact. Indirect transmission occurs when the organisms are transmitted from a source, either animate or inanimate, to a host by means of an inanimate object. An example is transmission of *Pseudomonas* organisms from one individual to another by means of a shaving brush. Droplet spread refers to organisms that travel through the air very short distances, that is, less than 3 feet from a source to a host. Therefore, the organisms are not airborne in the true sense. An example of a disease that may be spread by droplets is measles.

Common-vehicle transmission refers to agents transmitted by a common inanimate vehicle, with multiple cases resulting from such exposure. This category includes diseases in which food or water as well as drugs and parenteral fluids are the vehicles of infection. Examples include food-borne salmonellosis, waterborne shigellosis, and bacteremia resulting from use of intravenous fluids contaminated with a gram-negative organism.

The third method of transmission, airborne transmission, refers to infection spread by droplet nuclei or dust. To be truly airborne, the particles should travel more than 3 feet through the air from the source to the host. Droplet nuclei are the residue from the evaporation of fluid from droplets, are light enough to be transmitted more than 3 feet from the source, and may remain airborne for prolonged periods. Tuberculosis is primarily an airborne disease; the source may be a coughing patient who creates aerosols of droplet nuclei that contain tubercle bacilli. Infectious agents may be contained in dust particles, which may become resuspended and transmitted to hosts. An example occurred in an outbreak of salmonellosis in a newborn nursery in which *Salmonella*-contaminated dust in a vacuum cleaner bag was resuspended when the equipment was used repeatedly, resulting in infections among the newborns.

The fourth method of transmission is vector borne transmission, in which arthropods are the vectors. Vector transmission may be external or internal. External, or mechanical, transmission occurs when organisms are carried mechanically on the vector (for example, *Salmonella* organisms that contaminate the legs of flies). Internal transmission occurs when the organisms are carried within the vector. If the pathogen is not changed by its carriage within the vector, the carriage is called harborage (as when a flea ingests plague bacilli from an infected individual or animal and contaminates a susceptible host when it feeds again; the organism is not changed while in the flea). The other form of internal transmission is called biologic. In this form, the organism is changed biologically during its passage through the vector (for example, malaria parasites in the mosquito vector).

An infectious agent may be transmitted by more than one route. For example, *Salmonella* may be transmitted by a common vehicle (food) or by contact spread (human carrier). *Francisella tularensis* may be transmitted by any of the four routes.

Host

The third link in the chain of infection is the host. The organism may enter the host through the skin, mucous membranes, lungs, gastrointestinal tract, or genitourinary tract, and it may enter fetuses through the placenta. The resulting disease often reflects the point of entrance, but not always: meningococci that enter the host through the mucous membranes may nonetheless cause meningitis. Development of disease in a host reflects agent characteristics (see above) and is influenced by host defense mechanisms, which may be nonspecific or specific.

Nonspecific defense mechanisms include the skin, mucous membranes, secretions, excretions, enzymes, the inflammatory response, genetic factors, hormones, nutrition, behavioral patterns, and the presence of other diseases. Specific defense mechanisms or immunity may be natural, resulting from exposure to the infectious agent, or artificial, resulting from active or passive immunization.

The environment can affect any link in the chain of infection. Temperature can assist or inhibit multiplication of organisms at their reservoir; air velocity can assist the airborne movement of droplet nuclei; low humidity can damage mucous membranes; and ultraviolet radiation can kill the microorganisms. In any investigation of disease, it is important to evaluate the effect of environmental factors. At times, environmental control measures are instituted more on emotional grounds than on the basis of epidemiologic fact. It should be apparent that the occurrence of disease results from the interaction of many factors. Some of these factors are outlined here.

Epidemiologic Methods

The three major epidemiologic techniques are descriptive, analytic, and experimental. Although all three can be used in investigating the occurrence of disease, the method used most is descriptive epidemiology. Once the basic epidemiology of a disease has been described, specific analytic methods can be used to study the disease further, and a specific experimental approach can be developed to test a hypothesis.

Descriptive Epidemiology

In descriptive epidemiology, data that describe the occurrence of the disease are collected by various methods from all relevant sources. The data are then collated by time, place, and person. Four time trends are considered in describing the epidemiologic data. The secular trend describes the occurrence of disease over a prolonged period, usually years; it is influenced by the degree of immunity in the population and possibly nonspecific measures such as improved socioeconomic and nutritional levels among the population. For example, the secular trend of tetanus in the United States since 1920 shows a gradual and steady decline.

The second time trend is the periodic trend. A temporary modification in the overall secular trend, the periodic trend may indicate a change in the antigenic characteristics of the disease agent. For example, the change in antigenic structure of the prevalent influenza A virus every 2 to 3 years results in periodic increases in the occurrence of clinical influenza caused by lack of natural immunity among the population. Additionally, a lowering of the overall immunity of a population or a segment thereof (known as herd immunity) can result in an increase in the occurrence of the disease. This can be seen with some immunizable diseases when periodic decreases occur in the level of immunization in a defined population. This may then result in an increase in the number of cases, with a subsequent rise in the overall level of herd immunity. The number of new cases then decreases until the herd's immunity is low enough to allow transmission to occur again and new cases then appear.

The third time trend is the seasonal trend. This trend reflects seasonal changes in disease occurrence following changes in environmental conditions that enhance the ability of the agent to replicate or be transmitted. For example, food-borne disease outbreaks occur more frequently in the summer, when temperatures favor multiplication of bacteria. This trend becomes evident when the occurrence of salmonellosis is examined on a monthly basis.

The fourth time trend is the epidemic occurrence of disease. An epidemic is a sudden increase in occurrence due to prevalent factors that support transmission.

A description of epidemiologic data by place must consider three different sites: where the individual was when disease occurred; where the individual was when he or she became infected from the source; and where the source became infected with the etiologic agent. Therefore, in an outbreak of food poisoning, the host may become clinically ill at home from food eaten in a restaurant. The vehicle may have been undercooked chicken, which became infected on a poultry farm. These differences are important to consider in attempting to prevent additional cases.

The third focus of descriptive epidemiology is the infected person. All pertinent characteristics should be noted: age, sex, occupation, personal habits, socioeconomic status, immunization history, presence of underlying disease, and other data.

Once the descriptive epidemiologic data have been analyzed, the features of the epidemic should be clear enough that additional areas for investigation are apparent.

Analytic Epidemiology

The second epidemiologic method is analytic epidemiology, which analyzes disease determinants for possible causal relations. The two main analytic methods are the case-control (or case-comparison) method and the cohort method. The case-control method starts with the effect (disease) and retrospectively investigates the cause that led to the effect. The case group consists of individuals with the disease; a comparison group has members similar to those of the case group except for absence of the disease. These two groups are then compared to determine differences that would explain the occurrence of the disease. An example of a case-control study is selecting individuals with meningococcal meningitis and a comparison group matched for age, sex, socioeconomic status, and residence, but without the disease, to see what factors may have influenced the occurrence in the group that developed disease.

The second analytic approach is the cohort method, which prospectively studies two populations: one that has had contact with the suspected causal factor under study and a similar group that has had no contact with the factor. When both groups are observed, the effect of the factor should become apparent. An example of a cohort approach is to observe two similar groups of people, one composed of individuals who received blood transfusions and the other of persons who did not. The occurrence of hepatitis prospectively in both groups permits one to make an association between blood transfusions and hepatitis; that is, if the transfused blood was contaminated with hepatitis B virus, the recipient cohort should have a higher incidence of hepatitis than the nontransfused cohort.

The case-control approach is relatively easy to conduct, can be completed in a shorter period than the cohort approach, and is inexpensive and reproducible; however, bias may be introduced in selecting the two groups, it may be difficult to exclude subclinical cases from the comparison group, and a patient's recall of past events may be faulty. The advantages of a cohort study are the accuracy of collected data and the ability to make a direct estimate of the disease risk resulting from factor contact; however, cohort studies take longer and are more expensive to conduct.

Another analytic method is the cross-sectional study, in which a population is surveyed over a limited period to determine the relationship between a disease and variables present at the same time that may influence its occurrence.

Experimental Epidemiology

The third epidemiologic method is the experimental approach. A hypothesis is developed and an experimental model is constructed in which one or more selected factors are manipulated. The effect of the manipulation will either confirm or disprove the hypothesis. An example is the evaluation of the effect of a new drug on a disease. A group of people with the disease is identified, and some members are randomly selected to receive the drug. If the only difference between the two is use of the drug, the clinical differences between the groups should reflect the effectiveness of the drug.

Epidemic Investigation

An epidemic investigation describes the factors relevant to an outbreak of disease; once the circumstances related to the occurrence of disease are defined, appropriate control and prevention measures can be identified. In an epidemic investigation, data are collected, collated according to time, place, and person, and analyzed and inferences are drawn.

In the investigation, the first action should be to confirm the existence of the epidemic by noting from past surveillance data the number of cases suspected and comparing this with the number of cases initially reported. Additionally, the investigator should discuss the occurrence of the disease with physicians or others who have seen or reported cases after examining patients and reviewing laboratory and hospital records. These diagnoses should then be verified. A case definition should be developed to differentiate patients who represent actual cases, those who represent suspected or presumptive cases, and those who should be omitted from further study. Additional cases may be sought or additional patient data obtained, and a rough case count made.

This initial phase consists basically of collecting data, which then must be organized according to time, place, and person. The population at risk should be identified and a hypothesis developed concerning the occurrence of the disease. If appropriate, specimens should be collected and transported to the laboratory. More specific studies may be indicated. Additional data from these studies should be analyzed and the hypothesis confirmed or altered. After analysis, control and prevention measures should be developed and, as far as possible, implemented. A report containing this information should be prepared and distributed to those involved in investigating the outbreak and in implementing control and/or prevention measures. Continued surveillance activities may be appropriate to evaluate the effectiveness of the control and prevention measures.

In the United States, the CDC assists state health departments by providing epidemiologic and laboratory support services on request. Its assistance supports disease investigations and diagnostic laboratory activities and includes various training programs conducted in the states and at the CDC. A close working relationship exists between the CDC and state health departments. Additionally, physicians frequently consult with CDC personnel on a variety of health-related problems and attend public health training programs.

The use of epidemiology to characterize a disease before its etiology has been identified is exemplified by the initial studies of acquired immune deficiency syndrome (AIDS). The first cases came to the attention of the CDC late in 1981 when an increase was observed in requests for pentamidine for treatment of *Pneumocystis carinii* pneumonia. This initiated specific surveillance activities and epidemiologic studies that provided important information about this newly diagnosed disease.

Initial symptoms include fever, loss of appetite, weight loss, extreme fatigue, and enlargement of lymph nodes. A severe immune deficiency then develops, which appears to be associated with opportunistic infections. These infections include *P carinii* pneumonia, diagnosed in 52 percent of cases; Kaposi sarcoma in 26 percent of cases; and both *P carinii* pneumonia and Kaposi sarcoma in 7 percent of cases. The remaining 15 percent of AIDS patients have other parasitic, fungal, bacterial, or viral infections associated with immunodeficiencies. Among the first 2,640 cases reported to the CDC, there were 1,092 deaths, a case-fatality rate of 41 percent. Approximately 95 percent of the cases were male; 70 percent were 20 to 49 years of age at the time of diagnosis. Approximately 40 percent of the cases were reported from New York City, 12 percent from San Francisco, 8 percent from Los Angeles, and the remainder from 32 other states. Cases were reported from at least 16 other countries. Among the 90 percent of patients who were categorized according to possible risk factors, those at highest risk were homosexuals or bisexuals (70 percent), intravenous drug abusers (17 percent), Haitian entrants into the United States (9.5 percent), and persons with hemophilia (1 percent).

Analysis of these initial data, collected before the etiologic agent of AIDS was identified, supported the hypothesis that transmission occurred primarily by sexual contact, receipt of contaminated blood or blood products, or contact with contaminated intravenous needles. Spread through casual contact did not seem likely. The epidemiologic data indicated that AIDS was an infectious disease. It has now been determined that AIDS results from infection with a retrovirus of the human T cell leukemia/lymphoma virus family, which has been designated human immunodeficiency virus type I (HIV-l). The initial hypotheses have been proven as shown by analysis of data subsequently collected.

8

Principles of Diagnosis

INTRODUCTION

Some infectious diseases are distinctive enough to be identified clinically. Most pathogens, however, can cause a wide spectrum of clinical syndromes in humans. Conversely, a single clinical syndrome may result from infection with any one of many pathogens. Influenza virus infection, for example, causes a wide variety of respiratory syndromes that cannot be distinguished clinically from those caused by streptococci, mycoplasmas, or more than 100 other viruses.

Most often, therefore, it is necessary to use microbiologic laboratory methods to identify a specific etiologic agent. Diagnostic medical microbiology is the discipline that identifies etiologic agents of disease. The job of the clinical microbiology laboratory is to test specimens from patients for microorganisms that are, or may be, a cause of the illness and to provide information (when appropriate) about the in vitro activity of antimicrobial drugs against the microorganisms identified.

The staff of a clinical microbiology laboratory should be qualified to advise the physician as well as process specimens. The physician should supply salient information about the patient, such as age and sex, tentative diagnosis or details of the clinical syndrome, date of onset, significant exposures, prior antibiotic therapy, immunologic status, and underlying conditions. The clinical microbiologist participates in decisions regarding the microbiologic diagnostic studies to be performed, the type and timing of specimens to be collected, and the conditions for their transportation and storage. Above all, the clinical microbiology laboratory, whenever appropriate, should provide an interpretation of laboratory results.

Manifestations of Infection

The manifestations of an infection depend on many factors, including the site of acquisition or entry of the microorganism; organ or system tropisms of the microorganism; microbial virulence; the age, sex, and immunologic status of the patient; underlying diseases or conditions; and the presence of implanted prosthetic devices or materials. The signs and symptoms of infection may be localized, or they may be systemic, with fever, chills, and hypotension. In some instances the manifestations of an infection are sufficiently characteristic to suggest the diagnosis; however, they are often nonspecific.

Microbial Causes of Infection

Infections may be caused by bacteria (including mycobacteria, chlamydiae, mycoplasmas, and rickettsiae), viruses, fungi, or parasites. Infection may be endogenous or exogenous. In endogenous infections, the microorganism (usually a bacterium) is a component of the patient's indigenous flora. Endogenous infections can occur when the microorganism is aspirated from the upper to the lower respiratory tract or when it penetrates the skin or mucosal barrier as a result of trauma or surgery.

In contrast, in exogenous infections, the microorganism is acquired from the environment (e.g., from soil or water) or from another person or an animal. Although it is important to establish the cause of an infection, the differential diagnosis is based on a careful history, physical examination, and appropriate radiographic and laboratory studies, including the selection of appropriate specimens for microbiologic examination.

Results of the history, physical examination, and radiographic and laboratory studies allow the physician to request tests for the microorganisms most likely to be the cause of the infection.

Specimen Selection, Collection and Processing

Specimens selected for microbiologic examination should reflect the disease process and be collected in sufficient quantity to allow complete microbiologic examination. The number of microorganisms per milliliter of a body fluid or per gram of tissue is highly variable, ranging from less than 1 to 10^8 or 10^{10} colony-forming units (CFU). Swabs, although popular for specimen collection, frequently yield too small a specimen for accurate microbiologic examination and should be used only to collect material from the skin and mucous membranes.

Because skin and mucous membranes have a large and diverse indigenous flora, every effort must be made to minimize specimen contamination during collection. Contamination may be avoided by various means. The skin can be disinfected before aspirating or incising a lesion. Alternatively, the contaminated area may be bypassed altogether. Examples of such approaches are transtracheal puncture with aspiration of lower respiratory secretions or suprapubic bladder puncture with aspiration of urine. It is often impossible to collect an uncontaminated specimen, and decontamination procedures, cultures on selective media, or quantitative cultures must be used (see above).

Specimens collected by invasive techniques, particularly those obtained intraoperatively, require special attention. Enough tissue must be obtained for both histopathologic and microbiologic examination. Histopathologic examination is used to distinguish neoplastic from inflammatory lesions and acute from chronic inflammations. The type of inflammation present can guide the type of microbiologic examination performed. If, for example, a caseous granuloma is observed histopathologically, microbiologic examination should include cultures for mycobacteria and fungi.

The surgeon should obtain several samples for examination from a single large lesion or from each of several smaller lesions. If an abscess is found, the surgeon should collect several milliliters of pus, as well as a portion of the wall of the abscess, for microbiologic examination. Swabs should be kept out of the operating room.

If possible, specimens should be collected before the administration of antibiotics. Above all, close communication between the clinician and the microbiologist is essential to ensure that appropriate specimens are selected and collected and that they are appropriately examined.

MICROBIOLOGIC EXAMINATION

Direct Examination

Direct examination of specimens frequently provides the most rapid indication of microbial infection. A variety of microscopic, immunologic, and hybridization techniques have been developed for rapid diagnosis.

Sensitivity and Specificity

The sensitivity of a technique usually depends on the number of microorganisms in the specimen. Its specificity depends on how morphologically unique a specific microorganism appears microscopically or how specific the antibody or genetic probe is for that genus or species. For example, the sensitivity of Cram stains is such that the observation of two bacteria per oil immersion field (X 1,000) of a Gram-stained smear of uncentrifuged urine is equivalent to the presence of $> 10^5$ CFU/ml of urine. The sensitivity of the Gram-stained smear for detecting Gram-negative coccobacilli in cerebrospinal fluid from children with *Haemophilus influenzae* meningitis is approximately 75 percent because in some patients the number of colony-forming units per milliliter of cerebrospinal fluid is less than 10^4. At least 10^4 CFU of tubercle bacilli per milliliter of sputum must be present to be detected by an acid-fast smear of decontaminated and concentrated sputum.

An increase in the sensitivity of a test is often accompanied by a decrease in specificity. For example, examination of a Gram-stained smear of sputum from a patient with pneumococcal pneumonia is highly sensitive but also highly nonspecific if the criterion for defining a positive test is the presence of any Gram-positive cocci. If, however, a positive test is defined as the presence of a preponderance of Gram-positive, lancet-shaped diplococci, the test becomes highly specific but has a sensitivity of only about 50 percent. Similar problems related to the number of microorganisms present affect the sensitivity of immunoassays and genetic probes for bacteria, chlamydiae, fungi and viruses. In some instances, the sensitivity of direct examination tests can be improved by collecting a better specimen. For example, the sensitivity of fluorescent antibody stain for Chlamydia trachomatis is higher when endocervical cells are obtained with a cytobrush than with a swab. The sensitivity may also be affected by the stage of the disease at which the specimen is collected. For example, the detection of herpes simplex virus by immunofluorescence, immunoassay, or culture is highest when specimens from lesions in the vesicular stage of infection are examined. Finally, sensitivity may be improved through the use of an enrichment or enhancement step in which microbial or genetic replication occurs to the point at which a detection method can be applied.

Techniques

For microscopic examination it is sufficient to have a compound binocular microscope equipped with low-power (1OX), high-power (40X), and oil immersion (1OOX) achromatic objectives, 10X wide-field oculars, a mechanical stage, a substage condenser, and a good light source. For examination of wet-mount preparations, a darkfield condenser or condenser and objectives for phase contrast increases image contrast. An exciter barrier filter, darkfield condenser, and ultraviolet light source are required for fluorescence microscopy.

For immunologic detection of microbial antigens, latex particle agglutination, coagglutination, and enzyme-linked immunosorbent assay (ELISA) are the most frequently used techniques in the clinical laboratory. Antibody to a specific antigen is bound to latex particles or to a heat-killed and treated protein A-rich strain of Staphylococcus aureus to produce agglutination. There are several approaches to ELISA; the one most frequently used for the detection of microbial antigens uses an antigen-specific antibody that is fixed to a solid phase, which may be a latex or metal bead or the inside surface of a well in a plastic tray. The test is then completed by adding a second antigen-specific antibody bound to an enzyme that can react with a substrate to produce a colored product. When the enzyme-conjugated antibody is added, it binds to previously unbound antigenic sites, and the antigen is, in effect, sandwiched between the solid phase and the enzyme-conjugated antibody. The reaction is completed by adding the enzyme substrate.

Genetic probes are based on the detection of unique nucleotide sequences with the DNA or RNA of a microorganism. Once such a unique nucleotide sequence, which may represent a portion of a virulence gene or of chromosomal DNA, is found, it is isolated and inserted into a cloning vector (plasmid), which is then transformed into Escherichia coli to produce multiple copies of the probe. The sequence is then reisolated from plasmids and labeled with an isotope or substrate for diagnostic use. Hybridization of the sequence with a complementary sequence of DNA or RNA follows cleavage of the double-stranded DNA of the microorganism in the specimen.

The use of molecular technology in the diagnoses of infectious diseases has been further enhanced by the introduction of gene amplication techniques, such as the polymerase chain reaction (PCR) in which DNA polymerase is able to copy a strand of DNA by elongating complementary strands of DNA that have been initiated from a pair of closely spaced oligonucleotide primers. This approach has had major applications in the detection of infections due to microorganisms that are difficult to culture (e.g. the human immunodeficiency virus) or that have not as yet been successfully cultured (e.g. the Whipple's disease bacillus).

Culture

In many instances, the cause of an infection is confirmed by isolating and culturing microorganism either in artificial media or in a living host. Bacteria (including mycobacteria and mycoplasmas) and fungi are cultured in either liquid (broth) or on solid (agar) artificial media. Liquid media provide greater sensitivity for the isolation of small numbers of microorganisms; however, identification of mixed cultures growing in liquid media requires subculture onto solid media so that isolated colonies can be processed separately for identification. Growth in liquid media also cannot ordinarily be quantitated. Solid media, although somewhat less sensitive than liquid media, provide isolated colonies that can be quantified if necessary and identified. Some genera and species can be recognized on the basis of their colony morphologies.

In some instances one can take advantage of differential carbohydrate fermentation capabilities of microorganisms by incorporating one or more carbohydrates in the medium along with a suitable pH indicator. Such media are called differential media (e.g., eosin methylene blue or MacConkey agar) and are commonly used to isolate enteric bacilli. Different genera of the Enterobacteriaceae can then be presumptively identified by the color as well as the morphology of colonies.

Culture media can also be made selective by incorporating compounds such as antimicrobial agents that inhibit the indigenous flora while permitting growth of specific microorganisms resistant to these inhibitors. One such example is Thayer-Martin medium, which is used to isolate *Neisseria gonorrhoeae*. This medium contains vancomycin to inhibit Gram-positive bacteria, colistin to inhibit most Gram-negative bacilli, trimethoprim-sulfamethoxazole to inhibit *Proteus* species and other species that are not inhibited by colistin and anisomycin to inhibit fungi. The pathogenic *Neisseria* species, *N gonorrhoeae* and *N meningitidis*, are ordinarily resistant to the concentrations of these antimicrobial agents in the medium.

The number of bacteria in specimens may be used to define the presence of infection. For example, there may be small numbers ($< 10^3$ CFU/ml) of bacteria in clean-catch, midstream urine specimens from normal, healthy women; with a few exceptions, these represent bacteria that are indigenous to the urethra and periurethral region. Infection of the bladder (cystitis) or kidney (pyelone-phritis) is usually accompanied by bacteriuria of about $> 10^4$ CFU/ml. For this reason, quantitative cultures of urine must always be performed. For most other specimens a semiquantitative streak method over the agar surface is sufficient. For quantitative cultures, a specific volume of specimen is spread over the agar surface and the number of colonies per milliliter is estimated. For semiquantitative cultures, an unquantitated amount of specimen is applied to the agar and diluted by being streaked out from the inoculation site with a sterile bacteriologic loop. The amount of growth on the agar is then reported semiquantitatively as many, moderate, or few (or 3+, 2+, or 1+), depending on how far out from the inoculum site colonies appear. An organism that grows in all streaked areas would be reported as 3+.

Chlamydiae and viruses are cultured in cell culture systems, but virus isolation occasionally requires inoculation into animals, such as suckling mice, rabbits, guinea pigs, hamsters, or primates. Rickettsiae may be isolated with some difficulty and at some hazard to laboratory workers in animals or embryonated eggs. For this reason, rickettsial infection is usually diagnosed serologically. Some viruses, such as the hepatitis viruses, cannot be isolated in cell culture systems, so that diagnosis of hepatitis virus infection is based on the detection of hepatitis virus antigens or antibodies.

Cultures are generally incubated at 35 to 37°C in an atmosphere consisting of air, air supplemented with carbon dioxide (3 to 10 percent), reduced oxygen (microaerophilic conditions), or no oxygen (anaerobic conditions), depending upon requirements of the microorganism. Since clinical specimens from bacterial infections often contain aerobic, facultative anaerobic, and anaerobic bacteria, such specimens are usually inoculated into a variety of general purpose, differential, and selective media, which are then incubated under aerobic and anaerobic conditions.

Microbial Identification

Microbial growth in cultures is demonstrated by the appearance of turbidity, gas formation, or discrete colonies in broth; colonies on agar; cytopathic effects or inclusions in cell cultures; or detection of genus- or species-specific antigens or nucleotide sequences in the specimen, culture medium, or cell culture system.

Identification of bacteria (including mycobacteria) is based on growth characteristics (such as the time required for growth to appear or the atmosphere in which growth occurs), colony and microscopic morphology, and biochemical, physiologic, and, in some instances, antigenic or nucleotide sequence characteristics. The selection and number of tests for bacterial identification depend upon the category of bacteria present (aerobic versus anaerobic, Gram-positive versus Gram-negative, cocci versus bacilli) and the expertise of the microbiologist examining the culture. Gram-positive cocci that grow in air with or without added CO_2 may be identified by a relatively small number of tests. The identification of most Gram-negative bacilli is far more complex and often requires panels of 20 tests for determining biochemical and physiologic characteristics. The identification of filamentous fungi is based almost entirely on growth characteristics and colony and microscopic morphology. Identification of viruses is usually based on characteristic cytopathic effects in different cell cultures or on the detection of virus- or species-specific antigens or nucleotide sequences.

Interpretation of Culture Results

Some microorganisms, such as *Shigella dysenteriae, Mycobacterium tuberculosis, Coccidioides immitis*, and influenza virus, are always considered clinically significant. Others that ordinarily are harmless components of the indigenous flora of the skin and mucous membranes or that are common in the environment may or may not be clinically significant, depending on the specimen source from which they are isolated. For example, coagulase-negative staphylococci are normal inhabitants of the skin, gastrointestinal tract, vagina, urethra, and the upper respiratory tract (i.e., of the nares, oral cavity, and pharynx). Therefore, their isolation from superficial ulcers, wounds, and sputum cannot usually be interpreted as clinically significant.

They do, however, commonly cause infections associated with intravascular devices and implanted prosthetic materials. However, because intravascular devices penetrate the skin and since cultures of an implanted prosthetic device can be made only after incision, the role of coagulase-negative staphylococci in causing infection can usually be surmised only when the microorganism is isolated in large numbers from the surface of an intravascular device, from each of several sites surrounding an implanted prosthetic device, or, in the case of prosthetic valve endocarditis, from several separately collected blood samples. Another example, *Aspergillus fumigatus,* is widely distributed in nature, the hospital environment, and upper respiratory tract of healthy people but may cause fatal pulmonary infections in leukemia patients or in those who have undergone bone marrow transplantation. The isolation of *A fumigatus* from respiratory secretions is a nonspecific finding, and a definitive diagnosis of invasive aspergillosis requires histologic evidence of tissue invasion.

Physicians must also consider that the composition of microbial species on the skin and mucous membranes may be altered by disease, administration of antibiotics, endotracheal or gastric intubation, and the hospital environment. For example, potentially pathogenic bacteria can often be cultured from the pharynx of seriously ill, debilitated patients in the intensive care unit, but may not cause infection.

Serodiagnosis

Infection may be diagnosed by an antibody response to the infecting microorganism. This approach is especially useful when the suspected microbial agent either cannot be isolated in culture by any known method or can be isolated in culture only with great difficulty. The diagnosis of hepatitis virus and Epstein-Barr virus infections can be made only serologically, since neither can be isolated in any known cell culture system. Although human immunodeficiency virus type 1 (HIV-1) can be isolated in cell cultures, the technique is demanding and requires special containment facilities. HIV-1 infection is usually diagnosed by detection of antibodies to the virus.

The disadvantage of serology as a diagnostic tool is that there is usually a lag between the onset of infection and the development of antibodies to the infecting microorganism. Although IgM antibodies may appear relatively rapidly, it is usually necessary to obtain acute- and convalescent-phase serum samples to look for a rising titer of IgG antibodies to the suspected pathogen. In some instances the presence of a high antibody titer when the patient is initially seen is diagnostic; often, however, the high titer may reflect a past infection, and the current infection may have an entirely different cause. Another limitation on the use of serology as a diagnostic tool is that immunosuppressed patients may be unable to mount an antibody response.

Antimicrobial Susceptibility

The responsibility of the microbiology laboratory includes not only microbial detection and isolation but also the determination of microbial susceptibility to antimicrobial agents. Many bacteria, in particular, have unpredictable susceptibilities to antimicrobial agents, and their susceptibilities can be measured in vitro to help guide the selection of the most appropriate antimicrobial agent.

Antimicrobial susceptibility tests are performed by either disk diffusion or a dilution method. In the former, a standardized suspension of a particular microorganism is inoculated onto an agar surface to which paper disks containing various antimicrobial agents are applied. Following overnight incubation, any zone diameters of inhibition about the disks are measured and the results are reported as indicating susceptibility or resistance of the microorganism to each antimicrobial agent tested.

An alternative method is to dilute on a log2 scale each antimicrobial agent in broth to provide a range of concentrations and to inoculate each tube or, if a microplate is used, each well containing the antimicrobial agent in broth with a standardized suspension of the microorganism to be tested. The lowest concentration of antimicrobial agent that inhibits the growth of the microorganism is the minimal inhibitory concentration (MIC). The MIC and the zone diameter of inhibition are inversely correlated. In other words, the more susceptible the microorganism is to the antimicrobial agent, the lower the MIC and the larger the zone of inhibition. Conversely, the more resistant the microorganism, the higher the MIC and the smaller the zone of inhibition.

The term susceptible means that the microorganism is inhibited by a concentration of antimicrobial agent that can be attained in blood with the normally recommended dose of the antimicrobial agent and implies that an infection caused by this microorganism may be appropriately treated with the antimicrobial agent. The term resistant indicates that the microorganism is resistant to concentrations of the antimicrobial agent that can be attained with normal doses and implies that an infection caused by this microorganism could not be successfully treated with this antimicrobial agent.

9

Antimicrobial Chemotherapy

INTRODUCTION

The earliest evidence of successful chemotherapy is from ancient Peru, where the Indians used bark from the cinchona tree to treat malaria. Other substances were used in ancient China, and we now know that many of the poultices used by primitive peoples contained antibacterial and antifungal substances. Modern chemotherapy has been dated to the work of Paul Ehrlich in Germany, who sought systematically to discover effective agents to treat trypanosomiasis and syphilis. He discovered p-rosaniline, which has antitrypanosomal effects, and arsphenamine, which is effective against syphilis. Ehrlich postulated that it would be possible to find chemicals that were selectively toxic for parasites but not toxic to humans. This idea has been called the "magic bullet" concept. It had little success until the 1930s, when Gerhard Domagk discovered the protective effects of prontosil, the forerunner of sulfonamide. Ironically, penicillin G was discovered fortuitously in 1929 by Fleming, who did not initially appreciate the magnitude of his discovery. In 1939 Florey and colleagues at Oxford University again isolated penicillin.

In 1944 Waksman isolated streptomycin and subsequently found agents such as chloramphenicol, tetracyclines, and erythromycin in soil samples. By the 1960s, improvements in fermentation techniques and advances in medicinal chemistry permitted the synthesis of many new chemotherapeutic agents by molecular modification of existing compounds. Progress in the development of novel antibacterial agents has been great, but the development of effective, nontoxic antifungal and antiviral agents has been slow. Amphotericin B, isolated in the 1950s, remains an effective antifungal agent, although newer agents such as fluconazole are now widely used. Nucleoside analogs such as acyclovir have proved effective in the chemotherapy of selected viral infections.

Biochemical Basis of Antimicrobial Action

Bacterial cells grow and divide, replicating repeatedly to reach the large numbers present during an infection or on the surfaces of the body. To grow and divide, organisms must synthesize or take up many types of biomolecules. Antimicrobial agents interfere with specific processes that are essential for growth and/or division. They can be separated into groups such as inhibitors of bacterial and fungal cell walls, inhibitors of cytoplasmic membranes, inhibitors of nucleic acid synthesis, and inhibitors of ribosome function. Antimicrobial agents may be either bactericidal, killing the target bacterium or fungus, or bacteriostatic, inhibiting its growth. Bactericidal agents are more effective, but bacteriostatic agents can be extremely beneficial since they permit the normal defenses of the host to destroy the microorganisms.

INHIBITION OF BACTERIAL CELL WALL SYNTHESIS

As noted in earlier chapters, bacteria are classified as Gram-positive and Gram-negative organisms on the basis of staining characteristics. Gram-positive bacterial cell walls contain peptidoglycan and teichoic or teichuronic acid, and the bacterium may or may not be surrounded by a protein or polysaccharide envelope. Gram-negative bacterial cell walls contain peptidoglycan, lipopolysaccharide, lipoprotein, phospholipid, and protein. The critical attack site of anti-cell-wall agents is the peptidoglycan layer. This layer is essential for the survival of bacteria in hypotonic environments; loss or damage of this layer destroys the rigidity of the bacterial cell wall, resulting in death.

Peptidoglycan synthesis occurs in three stages. The first stage takes place in the cytoplasm, where the low-molecular-weight precursors UDP-GlcAc and UDP-MurNAc-L-Ala-D-Glu-meso-Dap-D-Ala-D-Ala are synthesized. A number of antimicrobial agents interfere with these early steps in cell wall biosynthesis. UTP and N-acetylglucosamine alpha-1-P are converted to UDP-N-acetylglucosamine, which is subsequently converted by the enzyme phosphoenolpyruvate: UDP-GIcNAc-3-enol-pyruvyltransferase.

Fosfomycins block this transfer by a direct nucleophilic attack on the enzyme. Because mammalian enzymes such as enolase, pyruvate kinase, carboxykinases, and the shikimate enlaces are not inhibited by these compounds, fosfomycins have no effect on the host. Three amino acids are added to the muramyl peptide to yield a tripeptide to which two more amino acids will be linked. The dipeptide D-alanyl-D-alanine is synthesized from two molecules of D-alanine by the enzyme D-alanyl-D-alanine synthetase. D-Alanine is produced from L-alanine by an alanine racemase. Cycloserine inhibits both alanine racemase and D-alanyl-D-alanine synthetase owing to the structural similarity of cycloserine and D-alanine and to the fact that cycloserine actually binds to the enzymes better than the D-alanine.

The second stage of cell wall synthesis is catalyzed by membrane-bound enzymes. The nonnucleotide portion of the precursor molecules previously made are transferred sequentially to a carrier in the cytoplasmic membrane. This carrier is a phosphorylated undecaprenyl alcohol. The lipid carrier functions as a point of attachment to the membrane for the precursors and allows for transport of the subunits across the hydrophobic interior of the cytoplasmic membrane to the outside surface. Bacitracin is a peptide antibiotic that specifically interacts with the pyrophosphate derivate of the undecaprenyl alcohol, preventing further transfer of the muramylpentapeptide from the precursor nucleotide to the nascent peptidoglycan.

The third stage of cell wall synthesis involves polymerization of the subunits and the attachment of nascent peptidoglycan to the cell wall. Polymerization occurs by transfer of the new peptidoglycan chain from its carrier in the membrane to the nonreducing N-acetylglucosamine of the new saccharide-peptide that is attached to the membrane. The new peptidoglycan is attached to preexisting cell wall peptidoglycan by a transpeptidase reaction that involves peptide chains in both polymers, one of which must possess a D-alanyl-D-alanine terminus. It is believed that the transpeptidase enzyme cleaves the peptide bond between two D-alanyl residues in the pentapeptide and become acylated via the carbonyl group of the penultimate D-alanine residue. This final reaction is inhibited by ß-lactam antibiotics.

These antibiotics contain a critical four-membered ring, which undergoes an acylation reaction with the transpeptidases that cross-link the polymers mentioned above. The ß-lactam antibiotics are the penicillins (penams), cephalosporins (including oxacephems and cephamycins), penems, thienamycins (carbapenems), and aztreonam (monobactams). The enzymes involved in this final process of cell wall formation are called penicillin-binding proteins since they were discovered by labeling with radioactive penicillin G. The enzymes are different in Gram-positive and Gram-negative bacteria and in anaerobic species. Differences in the penicillin-binding proteins explain, to some extent, differences in antibacterial activity of the ß-lactam antibiotics.

The penicillin-binding protein, to which a particular ß-lactam antibiotic binds, affects the morphologic response of the bacterium to the agent. For example, some antibiotics bind to a penicillin-binding protein that is involved in forming the septum between dividing cells; as a result, the bacteria continue to grow into long filaments, which eventually die. Binding to another penicillin-binding protein results in rapid lysis of a bacterium because the wall bulges and the bacterium bursts. ß-Lactams such as mecillinam (an amidino penicillin) do not bind to the penicillin-binding proteins of Gram-positive bacteria and therefore do not affect these bacteria. Aztreonam binds only to Gram-negative penicillin-binding proteins and does not inhibit Gram-positive or anaerobic species.

Vancomycin interrupts cell wall synthesis by forming a complex with the C-terminal D-alanine residues of peptidoglycan precursors. Complex formation at the outer surface of the cytoplasmic membrane prevents the transfer of the precursors from a lipid carrier to the growing peptidoglycan wall by transglycosidases.

Biochemical reactions in the cell wall catalyzed by transpeptidases and D,D-carboxypeptidases are also inhibited by vancomycin and other glycopeptide antimicrobials. Because of its large size and complex structure, vancomycin does not penetrate the outer membrane of gram-negative organisms. With resistance to beta-lactams increasing in frequency among staphylococci and enterococci, glycopeptides such as vancomycin remain important therapeutic agents against such bacteria.

Antibiotics that Affect the Function of Cytoplasmic Membranes

Bacterial Cytoplasmic Membranes

Biologic membranes are composed basically of lipid, protein, and lipoprotein. The cytoplasmic membrane acts as a diffusion barrier for water, ions, nutrients, and transport systems. Most workers now believe that membranes are a lipid matrix with globular proteins randomly distributed to penetrate through the lipid bilayer. A number of antimicrobial agents can cause disorganization of the membrane.

These agents can be divided into cationic, anionic, and neutral agents. The best-known compounds are polymyxin B and colistemethate (polymyxin E). These high-molecular-weight octapeptides inhibit Gram-negative bacteria that have negatively charged lipids at the surface. Since the activity of the polymyxins is antagonized by Mg^{2+} and Ca^{2+}, they probably competitively displace Mg^{2+} or Ca^{2+} from the negatively charged phosphate groups on membrane lipids. Basically, polymyxins disorganize membrane permeability so that nucleic acids and cations leak out and the cell dies.

The polymyxins are of virtually no use as systemic agents since they bind to various ligands in body tissues and are potent toxins for the kidney and nervous system. Gramicidins are also membrane-active antibiotics that appear to act by producing aqueous pores in the membranes. They also are used only topically.

Fungal Membranes

Fungal membranes contain sterols, whereas bacterial membranes do not. The polyene antibiotics, which apparently act by binding to membrane sterols, contain a rigid hydrophobic center and a flexible hydrophilic section. Structurally, polyenes are tightly packed rods held in rigid extension by the polyene portion. They interact with fungal cells to produce a membrane-polyene complex that alters the membrane permeability, resulting in internal acidification of the fungus with exchange of K+ and sugars; loss of phosphate esters, organic acids, nucleotides; and eventual leakage of cell protein. In effect, the polyene makes a pore in the fungal membrane and the contents of the fungus leak out. Prokaryotic cells neither bind to nor are inhibited by polyenes. Although numerous polyene antibiotics have been isolated, only amphotericin B is used systemically. Nystatin is used as a topical agent and primaricin as an ophthalmic preparation. A number of other agents interfere with the synthesis of fungal lipid membranes. These agents belong to a class of compounds referred to as imidazoles: miconazole, ketoconazole, clotrimazole, and fluconazole. These compounds inhibit the incorporation of subunits into ergosterol and may also directly damage the membrane.

Amphotericin B

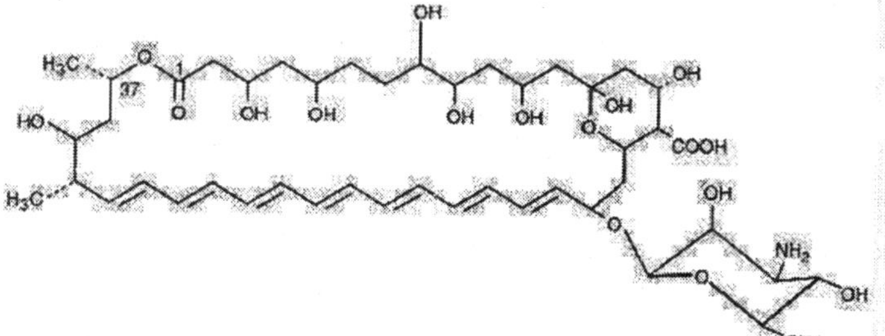

ANTIBIOTICS THAT INHIBIT NUCLEIC ACID SYNTHESIS

Antimicrobial agents can interfere with nucleic acid synthesis at several different levels. They can inhibit nucleotide synthesis or interconversion; they can prevent DNA from functioning as a proper template; and they can interfere with the polymerases involved in the replication and transcription of DNA.

Interference with Nucleotide Synthesis

A large number of agents interfere with purine and pyrimidine synthesis or with the interconversion or utilization of nucleotides. Other agents act as nucleotide analogs that are incorporated into polynucleotides.

Flucytosine (5-fluorocytosine) is an antifungal agent that inhibits yeast species. It is converted in the fungal cell to 5-fluorouracfl, which inhibits thymidylate synthetase resulting in a deficit of thymine nucleotides and impaired DNA synthesis. Adenosine arabinoside inhibits viruses. It is phosphorylated in virus-infected cells and acts as a competitive analog of DATP, inhibiting the incorporation of DATP into DNA. Acyclovir is a nucleoside analog that, after being converted to a triphosphate, inhibits the thymidine kinase and DNA polymerase of herpes viruses. Zidovudine (AZT) inhibits human immunodeficiency virus (HM replication by interfering with viral RNA-dependent DNA polymerase (reverse transcriptase).

Agents That Impair the Template Function of DNA

A number of substances bind to DNA by intercalation. None of them is useful as an antibacterial agent; however, chloroquine and miracil D (lucanthone) inhibit plasmodia and schistosomes, respectively. These agents are thought to intercalate into the DNA and thereby to inhibit further nucleic acid synthesis. Acridine dyes such as proflavine act by this intercalation mechanism, but because they are toxic and carcinogenic in mammals they cannot be used as antibacterial agents.

Inhibition of DNA-Directed DNA Polymerase

Rifamycins are a class of antibiotics that inhibit DNA-directed RNA polymerase (Fig. 11-6). Polypeptide chains in RNA polymerase attach to a factor that confers specificity for the recognition of promoter sites that initiate transcription of the DNA. Rifampin binds noncovalently but strongly to a subunit of RNA polymerase and interferes specifically with the initiation process. However, it has no effect once polymerization has begun.

Rifampin

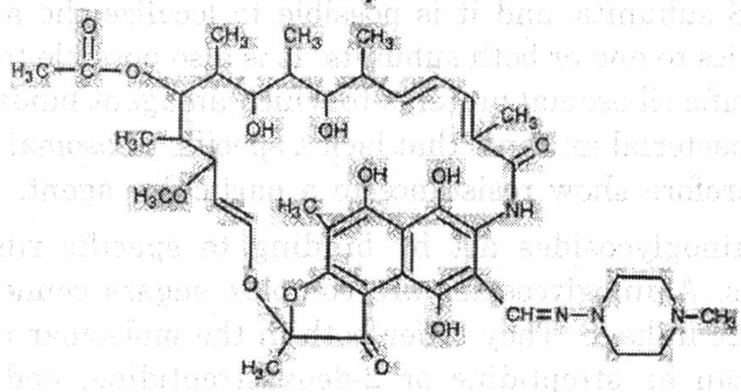

Inhibition of DNA Replication

DNA gyrase and topoisomerase I act in concert to maintain an optimum supercoiling state of DNA in the cell. In this capacity, DNA gyrase is essential for relieving torsional strain during replication of circular chromosomes in bacteria. The enzyme is a tetrameric protein composed of two A and two B subunits. A transient, covalent bond between the A subunit and DNA occurs during the double strand passage reaction catalyzed by gyrase.

Quinolones such as nalidixic acid, bind to the cleavage complex composed of DNA and gyrase during this strand passage. This interaction of quinolone acts to stabilize the cleavage intermediate which has a detrimental effect on the normal DNA replication process. The effects of this inhibition result in the death of the bacterial cell. The newer fluoroquinolones such as ciprofloxacin, norfloxacin, and ofloxacin also interact with DNA gyrase and possess a broad spectrum of antimicrobial activity.

Nitroimidazoles such as metronidazole inhibit anaerobic bacteria and protozoa. The nitro group of the nitrosohydroxyl amino moiety is reduced by an electron transport protein in anaerobic bacteria. The reduced drug causes strand breaks in the DNA. Mammalian cells are unharmed because they lack enzymes to reduce the nitro group of these agents.

Antimicrobial Inhibitors of Ribosome Function.

A number of antibacterial agents act by inhibiting ribosome function. Bacterial ribosomes contain two subunits, the 50S and 30S subunits, and it is possible to localize the action of antibiotics to one or both subunits. It is also possible to isolate the specific ribosomal proteins to which an agent binds and to isolate bacterial mutants that lack a specific ribosomal protein and therefore show resistance to a particular agent.

Aminoglycosides act by binding to specific ribosomal subunits. Aminoglycosides are complex sugars connected in glycosidic linkage. They differ both in the molecular nucleus, which can be streptidine or 2-deoxystreptidine, and in the aminohexoses linked to the nucleus. Essential to the activity of these agents are free NH, and OH groups by which aminoglycosides bind to specific ribosomal proteins. Streptomycin, the first aminoglycoside studied, was a useful too] in elucidating protein synthesis. However, it is rarely used clinically today except to treat tuberculosis, and its mode of action differs to some extent from that of the other clinically useful aminoglycosides, which are 2-deoxystreptidine derivatives such as gentamicin, tobramycin, and amikacin. Streptomycin binds to a specific S12 protein in the 30S ribosomal subunit and causes the ribosome to misread the genetic code.

Other aminoglycosides bind not only to the S12 protein of the 30S ribosome, but also to some extent to the L6 protein of the 50S ribosome. This latter binding is quite important in terms of the resistance of bacteria to aminoglycosides. Indeed, the aminoglycoside-type drugs can combine with other binding sites on 30S ribosomes, and they kill bacteria by inducing the formation of aberrant, nonfunctional complexes as well as by causing misreading.

Spectinomycin is an aminocylitol antibiotic that is closely related to the aminoglycosides. It binds to a different protein in the ribosome and is bacteriostatic but not bactericidal. It is used to treat penicillin-resistant gonorrhea.

Other agents that bind to 30S ribosomes are the tetracyclines. These agents appear to inhibit the binding of aminoacyl-tRNA into the A site of the bacteria] ribosome. Tetracycline binding is transient, so these agents are bacteriostatic. Nonetheless, they inhibit a wide variety of bacteria, chlamydias, and mycoplasmas and are extremely useful antibiotics.

There are three important classes of drugs that inhibit the 50S ribosomal subunit. Chloramphenicol is a bacteriostatic agent that inhibits both Gram-positive and Gram-negative bacteria. It inhibits peptide bond formation by binding to a peptidyltransferase enzyme on the 50S ribosome. Macrolides are large lactone ring compounds that bind to 50S ribosomes and appear to impair a peptidyltransferase reaction or translocation, or both. The most important macrolide is erythromycin, which inhibits Gram-positive species and a few Gram-negative species such as *Haemophilus, Mycoplasma, Chlamydia, and Legionella.*

New molecules such as azithromycin and clarithromycin have greater activity than erythromycin against many of these pathogens. Lincinoids, of which the most important is clindamycin, have a similar site of activity. Both macrolides and lincinoids are generally bacteriostatic. inhibiting only the formation of new peptide chains.

Drugs that Inhibit Other Biochemical Targets

Both trimethoprim and the sulfonamides interfere with folate metabolism in the bacterial cell by competitively blocking the biosynthesis of tetrahydrofolate, which acts as a carrier of one-carbon fragments and is necessary for the ultimate synthesis of DNA, RNA and bacterial cell wall proteins. Unlike mammals, bacteria and protozoan parasites usually lack a transport system to take up preformed folic acid from their environment. Most of these organisms must synthesize folates, although some are capable of using exogenous thymidine, circumventing the need for folate metabolism.

Sulfonamides competitively block the conversion of pteridine and p-aminobenzoic acid (PABA) to dihydrofolic acid by the enzyme pteridine synthetase. Sulfonamides have a greater affinity than p-aminobenzoic acid for pteridine synthetase. Trimethoprim has a tremendous affinity for bacterial dihydrofolate reductase (10,000 to 100,000 times higher than for the mammalian enzyme); when bound to this enzyme, it inhibits the synthesis of tetrahydrofolate.

Antibacterial Agents that Affect Mvcobacteria

Isoniazid is a nicotinamide derivative that inhibits mycobacteria. Its precise mode of action is not known, but it affects the synthesis of lipids, nucleic acids, and the mycolic acid of the cell walls of these species. Ethambutol is also an antimycobacterial agent whose mechanism of action is unknown. It is mycostatic, whereas isoniazid is mycocidal. The other antituberculosis drugs, rifampin and streptomycin, affect mycobacteria in the same manner that they inhibit bacteria. Pyrazinamide is a synthetic analog of nicotinamide. It is bactericidal, but its exact mechanism is unknown

Bacterial Resistance

Bacteria have proved adept at developing resistance to new antimicrobial agents. There are a number of ways in which bacteria can become resistant. Most of the early studies of bacterial resistance focused on single-step mutational events of chromosomal origin. Resistance to the early sulfonamides, for example, was the result of a single amino acid change in the enzyme pteridine synthetase that caused sulfonamides to bind less well than p-aminobenzoic acid. Similarly, a single step mutation that altered a ribosomal protein conferred resistance to streptomycin. In the late 1950s, Japanese workers found that enteric bacteria such as Shigella dysenteriae had become resistant not only to sulfonamides but also to the tetracyclines and chloramphenicol. This resistance was due not to a chromosomal change, but rather to the presence of extrachromosomal DNA that was transmissible. This type of resistance is called plasmid-mediated resistance.

Resistance-conferring plasmids are present in virtually all bacteria. For example, resistance to ampicillin appeared in *Haemophilus influenzae* in 1974 and in Neisseria gonorrhoeae in 1976. In the last several years, organisms such as enterococci have been shown to contain plasmids that confer resistance to drugs such as ampicillin and aminoglycosides.

Bacteria also contain transposons, which can insert into plasmids and also into the chromosome. Transposon-mediated resistance to most of the major antibiotics has been found in the past few years.

Antimicrobial agents exert a strong selective pressure on the development of both chromosomal and plasmid-mediated resistance, as discussed below. Administration of an antibiotic destroys the susceptible bacteria in a population, but may permit resistant ones to proliferate. From an epidemiologic viewpoint, plasmid-mediated resistance is the most important type, since it is transmissible, is usually highly stable, confers resistance to many different classes of antibiotics simultaneously, and often is associated with other characteristics that enable a microorganism to colonize and invade a susceptible host.

Mechanisms of Resistance

The basic mechanisms by which a microorganism can resist an antimicrobial agent are (1) to alter the receptor for the drug (the molecule on which it exerts its effect); (2) to decrease the amount of drug that reaches the receptor by altering entry or increasing removal of the drug; (3) to destroy or inactivate the drug; and (4) to develop resistant metabolic pathways. Bacteria can possess one or all of these mechanisms simultaneously.

Resistance Due to Altered Receptors

ß-Lactam Resistance. The ability to analyze changes in receptors for ß-lactams by competition experiments in which [^{14}C]penicillin is inhibited from binding to penicillin-binding proteins has explained a number of cases of bacterial resistance to penicillins and cephalosporins. In 1977, *Streptococcus pneumoniae* strains resistant to penicillin G were encountered in South Africa. Plasmids were not the cause of the resistance. Penicillin-resistant *S pneumoniae* cells have altered penicillin-binding proteins, which bind penicillin less well. Resistance of *S pneumoniae* to penicillin has been increasing, and there are now relatively resistant isolates (minimal inhibitory concentration [MIC], 0.1 to 1 mg/ml) in many parts of the world.

Altered penicillin-binding proteins also explain the resistance of some *Staphylococcus aureus* strains to ß-lactamase-stable penicillins (the so-called methicillin-resistant strains). The ß-lactams induce synthesis of a new penicillin-binding protein, PBP2a, which does not bind any ß-lactam. The ß-lactam resistance of coagulase-negative staphylococci is also the result of altered penicillin-binding proteins. Staphylococcal organisms resistant to methicillin are resistant to all penicillins, cephalosporins, and carbapenems.

The resistance of group D streptococci to ß-lactam antibiotics appears to be the result of lower affinity of the penicillin-binding proteins for the penicillins. Enterococci are resistant to all cephalosporins because of failure to bind to the penicillin-binding proteins. One Gram-negative species for which resistance to ß-lactam antibiotics can be correlated with diminished affinity of the target enzymes is *N gonorrhoeae*.

Vancomycin Resistance. Certain transposable genetic elements encode special cell wall-synthesizing enzymes which change the structure of the normal D-Ala-D-Ala side chain in the peptidoglycan assembly pathway. The altered side chain (D-Ala-D-Lac) does not bind vancomycin and allows normal peptidoglycan polymerization to occur in the presence of the drug. Depending upon the nature of the vancomycin resistance gene, high-level resistance can occur to glycopeptides. Thus far, this type of resistance has been found in enterococci but not in multi-resistant isolates of *Staphylococcus aureus*.

Macrolide-lincomycin Resistance. Macrolide-lincomycin resistance in clinical isolates of staphylococci and streptococci has been recognized for several decades. The resistance is due to methylation of two adenine nucleotides in the 23S component of 50S RNA. This resistance is plasmid mediated, and the resistance is encoded on transposons. Resistance results from induction of an enzyme that is normally repressed. The methylated RNA binds macrolidelincomycin-type drugs less well than unmethylated RNA does. Induction of resistance varies by species, and in most Gram-positive species erythromycin is a more effective inducer of resistance than is clindamycin. The plasmids that mediate macrolide-lincomycin resistance in streptococci and staphylococci have extensive structural similarity, indicating that these plasmids readily pass between these species.

Rifampin Resistance. The resistance of bacteria to rifampin is caused by an alternation of one amino acid in DNA-directed RNA polymerase, which results in reduced binding of rifampin. The degree of resistance is related to the degree to which the enzyme is changed, but does not correlate strictly with enzyme inhibition. This form of resistance occurs at a low level in any population of bacteria so that resistance develops by natural selection during a course of therapy. Naturally resistant organisms are more common among members of the Enterobacteriaceae, explaining why agents of urinary tract infections rapidly became resistant to rifampin. The resistance of *Neisseria meningitides* to rifampin appeared in closed military settings in which rifampin has been used for prophylaxis.

Sulfonamide-trimethoprim Resistance. Sulfonamide can be rendered ineffective by altered or new dihydropteroic synthetase that has poor affinity for sulfonamides and preferentially binds p-aminobenzoic acid. Sulfonamide resistance of this type can result from a point mutation or from acquisition of a plasmid that causes synthesis of the new enzyme. A most serious resistance problem is an increase in resistance to trimethoprim. This plasmid- and transposon-mediated resistance is due to production of an altered dihydrofolate reductase that has markedly reduced affinity for trimethoprim.

Quinolone Resistance. Resistance to quinolones can be caused by mutations in DNA gyrase subunits A or B, reduced outer membrane permeability in gram-negative cells, or to active efflux transporters found in many bacteria. The highest level of resistance to the newer fluoroquinolones is most frequently associated with chromosomal mutations, causing amino acid substitutions in a highly conserved region in the

A subunit of DNA gyrase. Multiple-mechanisms of resistance can occur in a single isolate of bacteria, leading to a higher level of resistance to many fluoroauinolones.

RESISTANCE DUE TO DECREASED ENTRY OF A DRUG

Tetracycline Resistance. The uptake of tetracycline by members of the Enterobacteriaceae is biphasic. In an initial energy-independent rapid phase, tetracycline binds to cell surface layers and passes by diffusion through the outer layers of the cell. In the second, energy-dependent phase, tetracycline crosses the cytoplasmic membrane, probably by means of a proton-motive force. The precise transport system has not been identified.

Tetracycline resistance is common in both Gram-positive and Gram-negative bacteria. In most cases it is plasmid encoded and inducible; however, chromosomal, constitutive resistance is found in some organisms such as *Proteus* species. Many plasmid-encoded specified tetracycline resistance determinants have been found in enteric bacteria. The most common of these determinants, TetB, is also present in *H influenzae*. Tetracycline resistance in *Staphylococcus aureus* is due primarily to small multicopy plasmids; chromosomal resistance is rare. Tetracycline resistance is found on nonconjugative plasmids in *Streptococcus faecalis* and on the chromosome of *S pneumoniae, S agalactiae* (group B streptococci), and oral streptococci. *Clostridium* species such as *C difficile* harbor chromosomal genes for tetracycline resistance.

Basically, tetracycline resistance is due to a decrease in the levels of drug accumulation. Decreased uptake and increased efflux both probably participate. Resistant bacteria bind less tetracycline, and the tetracycline they do accumulate is lost by an energy-dependent process when they are in a drug-free milieu.

Plasmid-mediated resistance to tetracyclines can be partially overcome in Gram-positive species by modifying the tetracycline nucleus. Hence, achievable concentrations of minocycline and doxycycline, in particular, will inhibit some tetracycline-resistant streptococci such as *S pneumoniae,* and some *S aureus* strains. Molecular modification has not been successful in overcoming the tetracycline resistance of members of the Enterobacteriaceae or *Pseudomonas* or most *Bacteroides* species.

Tetracycline resistance is a major concern because it is located on plasmids near insertion sites, and these plasmids readily acquire other genetic information to enlarge the spectrum of resistance. The widespread use of tetracycline in animal feeds may be a factor in the extensive, worldwide resistance of members of the Enterobacteriaceae, particularly enteric species such as *Salmonella,* to tetracyclines and subsequently to many other drugs. Not only can tetracycline resistance move among members of the Enterobacteriaceae on plasmids, but plasmids mediating tetracycline resistance have moved between *S aureus, S epidermidis, S pyogenes, S pneumoniae,* and *S faecalis.*

Fosfomycin Resistance. Fosfomycin and fosmidomycin, which inhibit cell wall synthesis, enter bacteria by means of a glycerol-phosphate or glucose-6-phosphate transport system. Gram-positive bacteria in which the glucose-6-phosphate transport system is poorly developed do not take up these drugs in concentrations adequate to inhibit the cell wall synthesis. This resistance usually is chromosomal. The resistance of Gram-negative bacteria to these agents is related primarily to the presence in the population of some bacteria that can function without the transport system. Plasmids and transposons that transfer resistance to fosfomycin have been found in bacteria such as *Serratia marcescens.*

Aminoglycoside Resistance. In the most important form of aminoglycoside resistance, the compound is modified outside the cell and resistance is due partly to poor uptake of the altered compound. Also, all aminoglycosides have free amino and hydroxy groups that are essential for binding to ribosomal proteins. A number of enzymes can acetylate the amino groups and phosphorylate or adenylate the hydroxyl groups. Other forms of resistance, such as altered binding site on 30S ribosomes, are much less common.

In members of the Enterobacteriaceae and in *Pseudomonas* species, the aminoglycosides pass through the cell wall via channels designed to admit cationic molecules to the periplasmic space. These channels, called porin channels, are lined by the porin protein.

Aminoglycosides are then translocated across the cell membrane by an energy-dependent proton-motive force and, in the cytoplasm, bind to ribosomes located just below the membrane. Aminoglycosides bind only to ribosomes actively engaged in protein synthesis. Binding to the ribosomes induces a protein involved in the uptake of the aminoglycosides.

Bacteria may contain in the periplasmic space enzymes that acetylate, phosphorylate, or adenylate aminoglycosides to various degrees. It is not clear whether the enzymes are free in the periplasmic space or bound to the cytoplasmic membrane. The modified aminoglycosides do not bind well to ribosomes, and hence uptake is poor or absent.

Aminoglycoside-modifying enzymes have been found in Gram-positive species such as *S aureus, S faecalis, S pyogenes,* and enzymes are particularly prevalent in members of the Enterobacteriaceae and *P aeruginosa. S pneumoniae.* These Many of the genes for aminoglycoside-modifying enzymes are carried on transposons

Anaerobic organisms such as *Bacteroides* species are resistant to aminoglycosides because they lack an oxygen-dependent transport system to move the drugs across the cytoplasmic membrane. Although most resistance of *S aureus* to aminoglycosides is due to aminoglycoside-modifying enzymes, small-colony variants of staphylococci also show resistance, which may be due to a defect in adenylate cyclase or in cyclic adenosine 5'-monophosphate (cAMP)binding proteins such that cells with a reduced growth rate do not transport aminoglycosides into the cytoplasm.

Some members of the Enterobacteriaceae and *Paeruginosa* appear to be resistant because of altered porin channels, since these bacteria do not take up any drug and do not have aminoglycoside-inactivating enzymes.

Resistance Due to Destruction or Inactivation of a Drug

Chloramphenicol Resistance. Many Gram-positive and Gram-negative bacteria, including some recently discovered *H influenzae* strains, are resistant to chloramphenicol because they possess the enzyme chloramphenicol transacetylate, which acetylates hydroxyl groups on the chloramphenicol structure. This enzyme, unlike the aminoglycoside-inactivating enzymes and ß-lactamases, is an intracellular enzyme of higher molecular weight and subunit structure. Acetylated chloramphenicol binds less well to the 50S ribosome.

ß-Lactam Resistance. The best-known mechanism of bacterial resistance is the resistance to ß-lactams, which is mediated by penicillinase enzymes. Resistance of *E coli* to penicillin was recognized in 1940, before sufficient penicillin was made to be clinically useful. In the 1940s, resistance of staphylococci was shown to be due to a penicillinase. As these enzymes also attack other ß-lactam compounds such as cephalosporins, carbapenems, and monobactams, they would be more appropriately designated ß-lactamases. The most important activity of these enzymes is alteration of the ß-lactam nucleus. ß-lactamases are widely distributed in nature and are usually classified on the basis of the principal compounds they destroy (e.g., as penicillinases or cephalosporinases). ß-Lactamases may be chromosomally or plasmid mediated, and they may be constitutive or inducible.

In Gram-positive species, ß-lactamases are primarily exoenzymes; that is, they are excreted into the milieu around the bacteria. Virtually all hospital isolates of staphylococci, both *S aureus* and *S epidermidis*, have ß-lactamases, and 50 to 80 percent of community-acquired staphylococcal isolates produce ß-lactamases. In Gram-negative species, both aerobic and anaerobic, ß-lactamases are contained in the periplasmic space, thus effectively protecting the penicillin-binding proteins.

Resistance of staphylococci to ß-lactams was soon overcome with the antistaphylococcal penicillins and the cephalosporins. Some strains of *S aureus* produce more ß-lactamase constitutively and can destroy some of the cephalosporins.

In 1974, *H influenzae* was shown to possess a plasmid mediated ß-lactamase. At present 10 to 35 percent of *H influenzae* strains in the United States produce ß-lactamases. The TnA transposon has become more widespread, and the resistance of *Haemophilus* species to penicillin G and ampicillin seems to be increasing yearly.

The *Haemophilus* ß-lactamase is the same structurally as the enzyme found in *E coli, Salmonella, Shigella,* and *N gonorrhoeae*. The enzyme has generally been called the TEM enzyme after the initials of the Greek girl from whom an *E coli* strain containing a plasmid ß-lactamase was first isolated. These enzymes are also called Richmond-Sykes class IIIa enzymes from a classification proposed by Richmond and Sykes in 1973.

By far the most common plasmid ß-lactamase found in nature is TEM-1, which accounts for 75 to 80 percent of plasmid-mediated ß-lactamase resistance worldwide. Recently new ß-lactamases have been found that hydrolyze compounds such as inomethoxy cephalosporin, which were not destroyed by other plasmid-encoded ß-lactamases. The new ß-lactamases have an altered amino acid composition, which permits binding to the cephalosporin and subsequent hydrolysis. How common these new enzymes will become is unknown.

Chromosomally mediated ß-lactamases are present in many *Enterobacter, Citrobacter, Proteus-Providencia,* and *Pseudomonas* species. All *Klebsiella* species possess a ß-lactamase, which acts primarily as a penicillinase and is chromosomally mediated. Constitutively produced ß-lactamases are also present in many anaerobic species.

SYNTHESIS OF RESISTANT METABOLIC PATHWAY

No synthesis of a new type of cell wall resistant to ß-lactams has occurred, but some bacteria, particularly some streptococci, lack the hydrolytic enzymes necessary for forming a new cell wall, and so ß-lactams do not lyse these bacteria. An altered hydrolytic system thus converts a bactericidal antibiotic into a bacteriostatic agent. Whether such resistance occurs in Gram-negative species is not clear.

Some thymidine-requiring streptococci are not inhibited by trimethoprim and sulfonamides and so are not killed by these agents. These organisms are a rare cause of urinary tract infections. Other bacteria produce adequate deoxyribosylthymine 5'-monophosphate (DTMP) by alternative methods and, as a result. survive exposure to these folate inhibitors.

Certain *Candida* or *Cryptococcus* yeasts are resistant to flucytosine because they cannot convert it to its active component, fluorouracil. Other fungi can resist the polyenes and imidazoles because they synthesize membrane components by different metabolic mechanisms.

Combinations of Antimicrobial Agents

Antibiotics are frequently used in combination for the following reasons: (1) to treat a life-threatening infection; (2) to prevent emergence of bacterial resistance; (3) to treat mixed infections of aerobic and anaerobic bacteria; (4) to enhance antibacterial activity (synergy); and (5) to use lower doses of a toxic drug. Combined treatment is reasonable when the precise agents of a serious infection are unknown. Use of two or more drugs to prevent the emergence of resistance is effective for tuberculosis and for therapy of some chronic infections. The use of combinations to achieve synergy is more complicated. Synergy occurs when a combination of two drugs causes inhibition or killing when used at a fourfold-lower concentration than that of either component drug used separately. However, indifference or antagonism may occur instead. Indifference means that the combined action is the same as with either component; antagonism refers to a reduction in the activity of one or both components in the presence of the other.

Important examples of bacterial synergy include (1) combinations of anti-cell wall agents with aminoglycosides, (2) use of ß-lactamase inhibitors with ß-lactamase-susceptible antibiotics, and (3) combinations of drugs that act on sequential steps in bacterial metabolic or synthetic pathways. Examples of synergy that have proved clinically important are the combination of penicillin and streptomycin to treat *Enterococcus faecalis* endocarditis and the combination of carbenicillin and gentamicin to treat P aeruginosa infections. Recently the ß-lactamase inhibitor clavulanic acid or sulbactam has been combined with amino penicillins to inhibit *S aureus, Klebsiella pneumoniae, H influenzae,* and anaerobic organisms such as *Bacteroides* species, all of which are resistant to amoxicillin or ampicillin when they contain ß-lactamases. The combination of sulfamethoxazole and trimethoprim attacks two parts of the folic acid cycle and synergistically inhibits many bacteria. Finally, combinations of two penicillins that affect different stages of cell wall synthesis in Gram-negative bacteria are synergistic. This is true for the combination of mecillinam (amdinocillin), which binds to PBP2 of bacteria, and penicillins or cephalosporins, which bind to PBP lb or 3.

Antagonism can occur when a bacteriostatic agent is combined with a bactericidal agent. The classic example has been the combination of chlortetracycline and penicillin in treatment of pneumococcal meningitis. This effect has not been explained from a molecular standpoint, and tetracyclines and penicillins or cephalosporins are used to treat mixed infections such as pelvic inflammatory disease due to *N gonorrhoeae* and *Chlamydia.* Some ß-lactam antibiotics can induce ß-lactamases that inactivate other ß-lactams, and antagonism can be shown in the test tube, but the relevance for clinical infections is not established.

Toxicology of Antimicrobial Agents

Antimicrobial agents can be directly toxic, can interact with other drugs to increase their toxicity, or can alter microbial flora to cause infection by organisms that are normally saprophytic. Allergic reactions can be caused by an agent, but penicillins can produce either immediate, IgE-mediated, or delayed hypersensitivity reactions. Cutaneous reactions have been reported with every class of antimicrobial agent. Hematologic reactions can range from the life-threatening blood dyscrasia that occurs in 1 in 60,000 individuals who receive chloramphenicol to hemolytic anemia due to sulfonamides in individuals who lack the enzyme glucose-6-phosphate dehydrogenase. Depression of blood platelet activity has occurred with many agents. By altering the gastrointestinal flora, almost all antibiotics can cause overgrowth of *Clostridium difficile*, which produces a toxin that causes diarrhea and even pseudomembranous colitis. Alteration of intestinal flora by antibiotics can also result in overgrowth of *Candida* in the mouth, vagina, or gastrointestinal tract. Since a number of antibiotics are metabolized in the liver, damage to the liver can occur. This has been of particular concern with isoniazid, which is used to treat tuberculosis. Damage to the kidneys can follow the use of aminoglycosides. Neurologic toxicity is fortunately fairly uncommon, but the aminoglycosides can damage the auditory or vestibular apparatus if the dosage is not closely monitored.

Bacteria continue to evolve new mechanisms of resistance to old and to new antimicrobial agents. Some bacteria such as *P aeruginosa* are particularly adept at utilizing a number of different mechanisms simultaneously to become resistant to agents in virtually every class and those with such diverse sites of action as cell wall, protein biosynthesis, or DNA and RNA synthesis. Progress in medicine will keep patients alive who have nosocomial infections with resistant pathogens.

Mechanism to Reduce Bacterial Resistance

Proper selection of new antibiotics will be a major force in slowing the development of antimicrobial resistance. Proper hygiene practices will reduce plasmid transfer and the establishment of multiple drug-resistant bacteria in the hospital and will delay the appearance of such species in the community.

10

Staphylococcus

INTRODUCTION

Bacteria in the genus Staphylococcus are pathogens of man and other mammals. Traditionally they were divided into two groups on the basis of their ability to clot blood plasma (the coagulase reaction). The coagulase-positive staphylococci constitute the most pathogenic species *S aureus*. The coagulase-negative staphylococci (CNS) are now known to comprise over 30 other species. The CNS are common commensals of skin, although some species can cause infections. It is now obvious that the division of staphylococci into coagulase positive and negative is artificial and indeed, misleading in some cases. Coagulase is a marker for *S aureus* but there is no direct evidence that it is a virulence factor. Also, some natural isolates of *S aureus* are defective in coagulase. Nevertheless, the term is still in widespread use among clinical microbiologists.

S aureus expresses a variety of extracellular proteins and polysaccharides, some of which are correlated with virulence. Virulence results from the combined effect of many factors expressed during infection. Antibodies will neutralize staphylococcal toxins and enzymes, but vaccines are not available. Both antibiotic treatment and surgical drainage are often necessary to cure abscesses, large boils and wound infections. Staphylococci are common causes of infections associated with indwelling medical devices. These are difficult to treat with antibiotics alone and often require removal of the device. Some strains that infect hospitalized patients are resistant to most of the antibiotics used to treat infections, vancomycin being the only remaining drug to which resistance has not developed.

Taxonomy

DNA-ribosomal RNA (rRNA) hybridization and comparative oligonucleotide analysis of 16S rRNA has demonstrated that staphylococci form a coherent group at the genus level. This group occurs within the broad Bacillus-Lactobacillus-Streptococcus cluster defining Gram-positive bacteria with a low G + C content of DNA.

At least 30 species of staphylococci have been recognized by biochemical analysis and in particular by DNA-DNA hybridization. Eleven of these can be isolated from humans as commensals. *S aureus* (nares) and *S epidermidis* (nares, skin) are common commensals and also have the greatest pathogenic potential. *S saprophyticus* (skin, occasionally) is also a common cause of urinary tract infection. *S haemolyticus, S simulans, S cohnii, S warneri* and *S lugdunensis* can also cause infections in man.

IDENTIFICATION OF STAPHYLOCOCCI IN THE CLINICAL LABORATORY

Structure

Staphylococci are Gram-positive cocci about 0.5 - 1.0 μm in diameter. They grow in clusters, pairs and occasionally in short chains. The clusters arise because staphylococci divide in two planes. The configuration of the cocci helps to distinguish micrococci and staphylococci from streptococci, which usually grow in chains. Observations must be made on cultures grown in broth, because streptococci grown on solid medium may appear as clumps. Several fields should be examined before deciding whether clumps or chains are present.

Catalase Test

The catalase test is important in distinguishing streptococci (catalase-negative) staphylococci which are catalase positive. The test is performed by flooding an agar slant or broth culture with several drops of 3% hydrogen peroxide. Catalase-positive cultures bubble at once. The test should not be done on blood agar because blood itself will produce bubbles.

Isolation and Identification

The presence of staphylococci in a lesion might first be suspected after examination of a direct Gram stain. However, small numbers of bacteria in blood preclude microscopic examination and require culturing first.

The organism is isolated by streaking material from the clinical specimen (or from a blood culture) onto solid media such as blood agar, tryptic soy agar or heart infusion agar. Specimens likely to be contaminated with other microorganisms can be plated on mannitol salt agar containing 7.5% sodium chloride, which allows the halo-tolerant staphylococci to grow. Ideally a Gram stain of the colony should be performed and tests made for catalase and coagulase production, allowing the coagulase-positive *S aureus* to be identified quickly. Another very useful test for *S aureus* is the production of thermostable deoxyribonuclease. *S aureus* can be confirmed by testing colonies for agglutination with latex particles coated with immunoglobulin G and fibrinogen which bind protein A and the clumping factor, respectively, on the bacterial cell surface. These are available from commercial suppliers (e.g., Staphaurex). The most recent latex test (Pastaurex) incorporates monoclonal antibodies to serotype 5 and 8 capsular polysaccharide in order to reduce the number of false negatives. (Some recent clinical isolates of *S aureus* lack production of coagulase and/or clumping factor, which can make identification difficult.)

The association of *S epidermidis* (and to a lesser extent of other coagulase-negative staphylococci) with nosocomial infections associated with indwelling devices means that isolation of these bacteria from blood is likely to be important and not due to chance contamination, particularly if successive blood cultures are positive. Nowadays, identification of S *epidermidis* and other species of *Staphylococcus* is performed using commercial biotype identification kits, such as API Staph Ident, API Staph-Trac, Vitek GPI Card and Microscan Pos

Combo. These comprise preformed strips containing test substrates.

Epidemiology of *Staphylococcus Aureus* Infections

Because *S aureus* is a major cause of nosocomial and community-acquired infections, it is necessary to determine the relatedness of isolates collected during the investigation of an outbreak. Typing systems must be reproducible, discriminatory, and easy to interpret and to use. The traditional method for typing *S aureus* is phage-typing. This method is based on a phenotypic marker with poor reproducibility. Also, it does not type many isolates (20% in a recent survey at the Center for Disease Control and Prevention), and it requires maintenance of a large number of phage stocks and propagating strains and consequently can be performed only by specialist reference laboratories.

Many molecular typing methods have been applied to the epidemiological analysis of *S aureus*, in particular, of methicillin-resistant strains (MRSA). Plasmid analysis has been used extensively with success, but suffers the disadvantage that plasmids can easily be lost and acquired and are thus inherently unreliable. Methods designed to recognize restriction fragment length polymorphisms (RFLP) using a variety of gene probes, including rRNA genes (ribotyping), have had limited success in the epidemiology of MRSA. In this technique the choice of restriction enzyme used to cleave the genomic DNA, as well as the probes, is crucial. Random primer PCR offers potential for discriminating between strains but a suitable primer has yet to be identified for *S aureus*. The method currently regarded as the most reliable is pulsed field gel electrophoresis, where genomic DNA is cut with a restriction enzyme that generates large fragments of 50-700 kb.

Clinical Manifestations of *S Aureus*

S aureus is notorious for causing boils, furuncles, styes, impetigo and other superficial skin infections in humans. It may also cause more serious infections, particularly in persons debilitated by chronic illness, traumatic injury, burns or immunosuppression. These infections include pneumonia, deep abscesses, osteomyelitis, endocarditis, phlebitis, mastitis and meningitis, and are often associated with hospitalized patients rather than healthy individuals in the community. *S aureus* and *S epidermidis* are common causes of infections associated with indwelling devices such as joint prostheses, cardiovascular devices and artificial heart valves.

Pathogenesis of *S Aureus* Infections

S aureus expresses many cell surface-associated and extracellular proteins that are potential virulence factors. For the majority of diseases caused by this organism, pathogenesis is multifactorial. Thus it is difficult to determine precisely the role of any given factor. This also reflects the inadequacies of many animal models for staphylococcal diseases.

However, there are correlations between strains isolated from particular diseases and expression of particular factors, which suggests their importance in pathogenesis. With some toxins, symptoms of a human disease can be reproduced in animals with pure proteins. The application of molecular biology has led to recent advances in the understanding of pathogenesis of staphylococcal diseases. Genes encoding potential virulence factors have been cloned and sequenced and proteins purified. This has facilitated studies at the molecular level on their modes of action, both in in vitro and in model systems. In addition, genes encoding putative virulence factors have been inactivated, and the virulence of the mutants compared to the wild-type strain in animal models. Any diminution in virulence implicates the missing factor. If virulence is restored when the gene is returned to the mutant then "Molecular Koch's Postulates" have been fulfilled. Several virulence factors of *S aureus* have been confirmed by this approach.

Adherence

In order to initiate infection the pathogen must gain access to the host and attach to host cells or tissues.

S aureus Adheres to Host Proteins

S aureus cells express on their surface proteins that promote attachment to host proteins such as laminin and fibronectin that form part of the extracellular matrix. Fibronectin is present on epithelial and endothelial surfaces as well as being a component of blood clots. In addition, most strains express a fibrinogen/fibrin binding protein (the clumping factor) which promotes attachment to blood clots and traumatized tissue. Most strains of *S aureus* express fibronectin and fibrinogen-binding proteins.

The receptor which promotes attachment to collagen is particularly associated with strains that cause osteomyelitis and septic arthritis. Interaction with collagen may also be important in promoting bacterial attachment to damaged tissue where the underlying layers have been exposed.

Evidence that these staphylococcal matrix-binding proteins are virulence factors has come from studying defective mutants in vitro adherence assays and in experimental infections. Mutants defective in binding to fibronectin and to fibrinogen have reduced virulence in a rat model for endocarditis, suggesting that bacterial attachment to the sterile vegetations caused by damaging the endothelial surface of the heart valve is promoted by fibronectin and fibrinogen.

Similarly, mutants lacking the collagen-binding protein have reduced virulence in a mouse model for septic arthritis. Furthermore, the soluble ligand-binding domain of the fibrinogen, fibronectin and collagen-binding proteins expressed by recombinant methods strongly blocks interactions of bacterial cells with the corresponding host protein.

Role of Adherence in Infections Associated with Medical Devices

Infections associated with indwelling medical devices ranging from simple intravenous catheters to prosthetic joints and replacement heart valves can be caused by *S aureus* and *S epidermidis*.

Very shortly after biomaterial is implanted in the human body it becomes coated with a complex mixture of host proteins and platelets. In one model system involving short-term contact between biomaterial and blood, fibrinogen was shown to be the dominant component and was primarily responsible for adherence of *S aureus* in subsequent in vitro assays. In contrast, with material that has been in the body for longer periods (e.g., human intravenous catheters) the fibrinogen is degraded and no longer promotes bacterial attachment. Instead, fibronectin, which remains intact, becomes the predominant ligand promoting attachment.

Adherence to Endothelial Cells

S aureus can adhere to the surface of cultured human endothelial cells and become internalized by a phagocytosis-like process. It is not clear if attachment involves a novel receptor or a known surface protein of *S aureus*. Some researchers think that *S aureus* can initiate endocarditis by attaching to the undamaged endothelium. Others feel that trauma of even a very minor nature is required to promote attachment of bacteria.

Avoidance of Host Defenses

S aureus expresses a number of factors that have the potential to interfere with host defense mechanisms. However, strong evidence for a role in virulence of these factors is lacking.

Capsular Polysaccharide

The majority of clinical isolates of *S aureus* express a surface polysaccharide of either serotype 5 or 8. This has been called a microcapsule because it can be visualized only by electron microscopy after antibody labeling, unlike the copious capsules of other bacteria which are visualized by light microscopy.

S aureus isolated from infections expresses high levels of polysaccharide but rapidly loses it upon laboratory subculture. The function of the capsule is not clear. It may impede phagocytosis, but in in vitro tests this was only demonstrated in the absence of complement.

Conversely, comparing wild-type and a capsule defective mutant strain in an endocarditis model suggested that polysaccharide expression actually impeded colonization of damaged heart valves, perhaps by masking adhesins.

Protein A

Protein A is a surface protein of *S aureus* which binds immunoglobulin G molecules by the Fc region. In serum, bacteria will bind IgG molecules the wrong way round by this non-immune mechanism. In principle this will disrupt opsonization and phagocytosis. Indeed mutants of *S aureus* lacking protein A are more efficiently phagocytozed in vitro, and studies with mutants in infection models suggest that protein A enhances virulence.

Leukocidin

S aureus can express a toxin that specifically acts on polymorphonuclear leukocytes. Phagocytosis is an important defense against staphylococcal infection so leukocidin should be a virulence factor. This toxin is discussed in more detail in the next section.

Damage to the Host

S aureus can express several different types of protein toxins which are probably responsible for symptoms during infections. Some damage the membranes of erythrocytes, causing hemolysis; but it is unlikely that hemolysis is relevant in vivo. The leukocidin causes membrane damage to leukocytes and is not hemolytic. Systemic release of a-toxin causes septic shock, while enterotoxins and TSST-1 cause toxic shock.

MEMBRANE DAMAGING TOXINS

(a) a-toxin

The best characterized and most potent membrane-damaging toxin of *S aureus* is a-toxin. It is expressed as a monomer that binds to the membrane of susceptible cells. Subunits then oligomerize to form hexameric rings with a central pore through which cellular contents leak.

Susceptible cells have a specific receptor for a-toxin which allows low concentrations of toxin to bind, causing small pores through which monovalent cations can pass. At higher concentrations, the toxin reacts non-specifically with membrane lipids, causing larger pores through which divalent cations and small molecules can pass. However, it is doubtful if this is relevant under normal physiological conditions.

In humans, platelets and monocytes are particularly sensitive to a-toxin. They carry high affinity sites which allow toxin to bind at concentrations that are physiologically relevant. A complex series of secondary reactions ensue, causing release of eicosanoids and cytokines which trigger production of inflammatory mediators. These events cause the symptoms of septic shock that occur during severe infections caused by *S aureus*.

The notion that a-toxin is a major virulence factor of *S aureus* is supported by studies with the purified toxin in animals and in organ culture. Also, mutants lacking a-toxin are less virulent in a variety of animal infection models.

(b) ß-toxin

ß-toxin is a sphingomyelinase which damages membranes rich in this lipid. The classical test for ß-toxin is lysis of sheep erythrocytes. The majority of human isolates of *S aureus* do not express ß-toxin. A lysogenic bacteriophage is inserted into the gene that encodes the toxin. This phenomenon is called negative phage conversion. Some of the phages that inactivate the ß-toxin gene carry the determinant for an enterotoxin and staphylokinase.

In contrast the majority of isolates from bovine mastitis express ß-toxin, suggesting that the toxin is important in the pathogenesis of mastitis. This is supported by the fact that ß-toxin-deficient mutants have reduced virulence in a mouse model for mastitis.

(c) d-toxin

The d-toxin is a very small peptide toxin produced by most strains of *S aureus*. It is also produced by *S epidermidis* and *S lugdunensis*. The role of d-toxin in disease is unknown.

(d) g-toxin and leukocidin

The g-toxin and the leukocidins are two-component protein toxins that damage membranes of susceptible cells. The proteins are expressed separately but act together to damage membranes. There is no evidence that they form multimers prior to insertion into membranes. The g-toxin locus expresses three proteins. The B and C components form a leukotoxin with poor hemolytic activity, whereas the A and B components are hemolytic and weakly leukotoxic.

The classical Panton and Valentine (PV) leukocidin is distinct from the leukotoxin expressed by the g-toxin locus. It has potent leukotoxicity and, in contrast to g-toxin, is non-hemolytic. Only a small fraction of *S aureus* isolates (2% in one survey) express the PV leukocidin, whereas 90% of those isolated from severe dermonecrotic lesions express this toxin. This suggests that PV leukocidin is an important factor in necrotizing skin infections.

PV-leukocidin causes dermonecrosis when injected subcutaneously in rabbits. Furthermore, at a concentration below that causing membrane damage, the toxin releases inflammatory mediators from human neutrophils, leading to degranulation. This could account for the histology of dermonecrotic infections (vasodilation, infiltration and central necrosis).

Superantigens: enterotoxins and toxic shock syndrome toxin

S aureus can express two different types of toxin with superantigen activity, enterotoxins, of which there are six serotypes (A, B, C, D, E and G) and toxic shock syndrome toxin (TSST-1). Enterotoxins cause diarrhea and vomiting when ingested and are responsible for staphylococcal food poisoning. When expressed systemically, enterotoxins can cause toxic shock syndrome (TSS) - indeed enterotoxins B and C cause 50% of non-menstrual TSS. TSST-1 is very weakly related to enterotoxins and does not have emetic activity. TSST-1 is responsible for 75% of TSS, including all menstrual cases. TSS can occur as a sequel to any staphylococcal infection if an enterotoxin or TSST-1 is released systemically and the host lacks appropriate neutralizing antibodies. Tampon-associated TSS is not a true infection, being caused by growth of *S aureus* in a tampon and absorption of the toxin into the blood stream.

TSS came to prominence with the introduction of super-absorbent tampons; and although the number of such cases has decreased dramatically, they still occur despite withdrawal of certain types of tampons from the market.

Superantigens stimulate T cells non-specifically without normal antigenic recognition. Up to one in five T cells may be activated, whereas only 1 in 10,000 are stimulated during antigen presentation. Cytokines are released in large amounts, causing the symptoms of TSS. Superantigens bind directly to class II major histocompatibility complexes of antigen-presenting cells outside the conventional antigen-binding grove. This complex recognizes only the Vb element of the T cell receptor. Thus any T cell with the appropriate Vb element can be stimulated, whereas normally antigen specificity is also required in binding.

Epidermolytic (exfoliative) toxin (ET)

This toxin causes the scalded skin syndrome in neonates, with widespread blistering and loss of the epidermis. There are two antigenically distinct forms of the toxin, ETA and ETB. There is evidence that these toxins have protease activity. Both toxins have a sequence similarity with the *S aureus* serine protease, and the three most important amino acids in the active site of the protease are conserved. Furthermore, changing the active site of serine to a glycine completely eliminated toxin activity. However, ETs do not have discernible proteolytic activity but they do have esterase activity. It is not clear how the latter causes epidermal splitting. It is possible that the toxins target a very specific protein which is involved in maintaining the integrity of the epidermis.

OTHER EXTRACELLULAR PROTEINS

Coagulase

Coagulase is not an enzyme. It is an extracellular protein which binds to prothrombin in the host to form a complex called staphylothrombin. The protease activity characteristic of thrombin is activated in the complex, resulting in the conversion of fibrinogen to fibrin. This is the basis of the tube coagulase test, in which a clot is formed in plasma after incubation with the *S aureus* broth-culture supernatant. Coagulase is a traditional marker for identifying *S aureus* in the clinical microbiology laboratory. However, there is no evidence that it is a virulence factor, although it is reasonable to speculate that the bacteria could protect themselves from host defenses by causing localized clotting. Notably, coagulase deficient mutants have been tested in several infection models but no differences from the parent strain were observed.

There is some confusion in the literature concerning coagulase and clumping factor, the fibrinogen-binding determinant on the *S aureus* cell surface. This is partly due to loose terminology, with the clumping factor sometimes being referred to as bound coagulase. Also, although coagulase is regarded as an extracellular protein, a small fraction is tightly bound on the bacterial cell surface where it can react with prothrombin. Finally, it has recently been shown that the coagulase can bind fibrinogen as well as thrombin, at least when it is extracellular. Genetic studies have shown unequivocally that coagulase and clumping factor are distinct entities. Specific mutants lacking coagulase retain clumping factor activity, while clumping factor mutants express coagulase normally.

Staphylokinase

Many strains of *S aureus* express a plasminogen activator called staphylokinase. The genetic determinant is associated with lysogenic bacteriophages. A complex formed between staphylokinase and plasminogen activates plasmin-like proteolytic activity which causes dissolution of fibrin clots. The mechanism is identical to streptokinase, which is used in medicine to treat patients suffering from coronary thrombosis. As with coagulase there is no evidence that staphylokinase is a virulence factor, although it seems reasonable to imagine that localized fibrinolysis might aid in bacterial spreading.

Enzymes

S aureus can express proteases, a lipase, a deoxyribonuclease (DNase) and a fatty acid modifying enzyme

(FAME). The first three probably provide nutrients for the bacteria, and it is unlikely that they have anything but a minor role in pathogenesis. However, the FAME enzyme may be important in abscesses, where it could modify anti-bacterial lipids and prolong bacterial survival. The thermostable DNase is an important diagnostic test for identification of *S aureus*.

Coagulase Negative Staphylococci

Staphylococci other than *S aureus* can cause infections in man. *S epidermidis* is the most important coagulase-negative staphylococcus (CNS) species and is the major cause of infections associated with prosthetic devices and catheters. CNS also cause peritonitis in patients receiving continuous ambulatory peritoneal dialysis and endocarditis in those with prosthetic valves. These infections are not usually nosocomially acquired. Other species such as *S haemolyticus, S warneri, S hominis, S capitis, S intermedius, S schleiferi* and *S simulans* are infrequent pathogens. *S lugdunesis* is a newly recognized species. It is probably more pathogenic than are other CNS species, with cases of endocarditis and other infections being reported. It is likely that the incidence of infections caused by these organisms is underestimated because of difficulties in identification.

Diagnosis of CNS infections is difficult. Infections are often indolent and chronic with few obvious symptoms. This is due to the smaller array of virulence factors and toxins compared to those in the case of *S aureus*. *S epidermidis* is a skin commensal and is one of the most common contaminants of samples sent to the diagnostic laboratory, while *S lugdunensis* is often confused with *S aureus*. Precise identification of CNS species requires the use of expensive test kits, such as the API-Staph.

In contrast to *S aureus*, little is known about mechanisms of pathogenesis of *S epidermidis* infections. Adherence is obviously a crucial step in the initiation of foreign body infections. Much research has been done on the interaction between *S epidermidis* and plastic material used in implants, and a polysaccharide adhesion (PS/A) has been identified.

Mutants lacking PS/A are less virulent in an animal model for foreign body infection, and immunization with purified PS/A is protective. Bacteria-plastic interactions are probably important in colonization of catheters through the point of entry. However, host proteins are quickly deposited on implants. *S epidermidis* does not bind to fibrinogen but most isolates bind fibronectin, albeit less avidly than *S aureus*. However, it is not known if a protein analogous to the fibronectin binding protein of *S aureus* is involved.

A characteristic of clinical isolates of *S epidermidis* is the production of "slime." This is a controversial topic. Some feel that slime is an in vitro manifestation of the ability to form a biofilm in vivo, for example on the surface of a prosthetic device, and is thus a virulence marker. In vitro, slime is formed during growth in broth as a biofilm on the surface of the growth vessel. The composition of this slime is probably influenced by the growth medium. One study with defined medium showed that the slime was predominantly secreted teichoic acid, a polymer normally found in the cell wall of staphylococci. Some polysaccharides in slime from bacteria grown on solid medium are derived from the agar.

Resistance of Staphylococci to Antimicrobial Drugs

Hospital strains of *S aureus* are often resistant to many different antibiotics. Indeed strains resistant to all clinically useful drugs, apart from the glycopeptides vancomycin and teicoplanin, have been described. The term MRSA refers to methicillin resistance and most methicillin-resistant strains are also multiply resistant.

Plasmid-associated vancomycin resistance has been detected in some enterococci and the resistance determinant has been transferred from enterococci to *S aureus* in the laboratory and may occur naturally. *S epidermidis* nosocomial isolates are also often resistant to several antibiotics including methicillin. In addition, *S aureus* expresses resistance to antiseptics and disinfectants, such as quaternary ammonium compounds, which may aid its survival in the hospital environment.

Since the beginning of the antibiotic era *S aureus* has responded to the introduction of new drugs by rapidly acquiring resistance by a variety of genetic mechanisms including (1) acquisition of extrachromosomal plasmids or additional genetic information in the chromosome via transposons or other types of DNA insertion and (2) by mutations in chromosomal genes.

Many plasmid-encoded determinants have recently become inserted into the chromosome at a site associated with the methicillin resistance determinant. There may be an advantage to the organism having resistance determinants in the chromosome because they will be more stable.There are essentially four mechanisms of resistance to antibiotics in bacteria: (1) enzymatic inactivation of the drug, (2) alterations to the drug target to prevent binding, (3) accelerated drug efflux to prevent toxic concentrations accumulating in the cell, and (4) a by-pass mechanism whereby an alternative drug-resistant version of the target is expressed.

FUTURE PROSPECTS

Antimicrobial Drugs

Ever since the first use of penicillin, *S aureus* has shown a remarkable ability to adapt. Resistance has developed to new drugs within a short time of their introduction. Some strains are now resistant to most conventional antibiotics. It is worrisome that there do not seem to be any new antibiotics on the horizon. Any recent developments have been modifications to existing drugs.

The original strategy used by the pharmaceutical industry to find antimicrobial drugs was to screen natural products and synthetic chemicals for antimicrobial activity. The mechanism of action was then investigated.

New approaches are being adopted to find the next generation of antimicrobials. Potential targets such as enzymes involved in an essential function (e.g., in cell division) are identified based on knowledge of bacterial physiology and metabolism. Screening methods are then developed to identify inhibitors of a specific target molecule. In addition, with detailed molecular knowledge of the target molecule, specific inhibitors can be designed.

Vaccines and New Approaches to Combatting Nosocomial Infections

No vaccine is currently available to combat staphylococcal infections. There may now be a case for considering methods to prevent disease, particularly in hospitalized patients.

Hyperimmune serum from human volunteer donors or humanized monoclonal antibodies directed towards surface components (e.g., capsular polysaccharide or surface protein adhesions) could both prevent bacterial adherence and also promote phagocytosis of bacterial cells. Indeed a prototype vaccine based on capsular polysaccharide from *S aureus* has been administered to volunteers to raise hyperimmune serum, which could be given to patients in hospital before surgery. A vaccine based on fibronectin binding protein induces protective immunity against mastitis in cattle and might also be used as a vaccine in humans.

When the molecular basis of the interactions between the bacterial surface proteins and the host matrix protein ligands are known it might be possible to design compounds that block the interactions and thus prevent bacterial colonization. These could be administered systemically or topically.

11

Streptococcus

INTRODUCTION

The genus *Streptococcus,* a heterogeneous group of Gram-positive bacteria, has broad significance in medicine and industry. Various streptococci are important ecologically as part of the normal microbial flora of animals and humans; some can also cause diseases that range from subacute to acute or even chronic. Among the significant human diseases attributable to streptococci are scarlet fever, rheumatic heart disease, glomerulonephritis, and pneumococcal pneumonia. Streptococci are essential in industrial and dairy processes and as indicators of pollution.

The nomenclature for streptococci, especially the nomenclature in medical use, has been based largely on serogroup identification of cell wall components rather than on species names. For several decades, interest has focused on two major species that cause severe infections: *S pyogenes* (group A streptococci) and *S pneumoniae* (pneumococci). In 1984, two members were assigned a new genus - the group D enterococcal species (which account for 98% of human enterococcal infections) became *Enterococcus faecalis* (the majority of human clinical isolates) and *E faecium* (associated with a remarkable capacity for antibiotic resistance).

In recent years, increasing attention has been given to other streptococcal species, partly because innovations in serogrouping methods have led to advances in understanding the pathogenetic and epidemiologic significance of these species.

A variety of cell-associated and extracellular products are produced by streptococci, but their cause-effect relationship with pathogenesis has not been defined. Some of the other medically important streptococci are *S agalactiae* (group B), an etiologic agent of neonatal disease; *E faecalis* (group D), a major cause of endocarditis, and the viridans streptococci. Particularly for the viridans streptococci, taxonomy and nomenclature are not yet fully reliable or consistent. Important members of the viridans streptococci, normal commensals, include *S mutans* and *S sanguis* (involved in dental caries), *S mitis* (associated with bacteremia, meningitis, periodontal disease and pneumonia), and "*S milleri*" (associated with suppurative infections in children and adults). There remains persistent taxonomic confusion regarding "*S milleri*."

Clinical Manifestations

In humans, diseases associated with the streptococci occur chiefly in the respiratory tract, bloodstream, or as skin infections. Human disease is most commonly associated with Group A streptococci. Acute group A streptococcal disease is most often a respiratory infection (pharyngitis or tonsillitis) or a skin infection (pyoderma). Also medically significant are the late immunologic sequelae, not directly attributable to dissemination of bacteria, of group A infections (rheumatic fever following respiratory infection and glomerulonephritis following respiratory or skin infection) which remain a major worldwide health concern.

Much effort is being directed toward clarifying the risk and mechanisms of these sequelae and identifying rheumatogenic and nephritogenic strains. *S pneumoniae* remains a primary cause of serious focal and systemic infections, the first most common cause of community acquired pneumonia in the United States and of fatal bacterial pneumonia in developing countries. Hemorrhagic shock in association with *S pneumoniae* sepsis in previously healthy children has been reported recently in the United States. Of major biologic importance is a renewed interest in safe and effective streptococcal vaccines.

Structure

Both *S pyogenes* and *S pneumoniae* are Gram-positive cocci, nonmotile, and nonsporulating; they usually require complex culture media. *S pyogenes* characteristically is a round-to-ovoid coccus 0.6-1.0 μm in diameter. They divide in one plane and thus occur in pairs, or (especially in liquid media or clinical material) in chains of varying lengths. S *pneumoniae* appears as a 0.5-1.25 μm diplococcus, typically described as lancet-shaped but sometimes difficult to distinguish morphologically from other streptococci. Streptococcal cultures older than the logarithmic phase, which is the most active growth period of a culture, may lose their Gram-positive staining characteristics.

Unlike *Staphylococcus*, all streptococci lack the enzyme catalase. Most are facultative anaerobes but some are obligate anaerobes. Streptococci often have a mucoid or smooth colonial morphology, and *S pneumoniae* colonies exhibit a central depression caused by rapid partial autolysis. As *S pneumoniae* colonies age, viability is lost during fermentative growth in the absence of catalase and peroxidase because of the accumulation of peroxide. Some group B and D streptococci produce pigment. Recently, nutritionally deficient streptococci (also known as wall-deficient, L form, thiol-requiring, satelliting, or pyridoxal-dependent) have been recovered from a variety of clinical sources, including blood, abscesses, and oral and urethral ulcers. These variants demonstrate bizarre pleomorphism microscopically and do not grow on routine subculture.

Classification, Antigenic Types and Extracellular Growth Products

The type of hemolytic reaction displayed on blood agar has long been used to classify the streptococci. b-Hemolysis is associated with complete lysis of red cells surrounding the colony, whereas a-hemolysis is a partial or "greeni. g" hemolysis associated with reduction of red cell hemoglobin. Nonhemolytic colonies have been termed g-hemolytic. Hemolysis is affected by the species and age of red cells as well as by other properties of the base medium. Use of the hemolytic reaction in classification is not completely satisfactory. Some group A streptococci appear nonhemolytic; group B can manifest a-, b-, or even g-hemolysis; most *S pneumoniae* are a-hemolytic but can cause ß-hemolysis during anaerobic incubation. The viridans group, although linked by the property of a-hemolysis, is actually an extremely diverse group of organisms that does not usually react with Lancefield grouping sera. The taxonomy and biochemical and genetic relationships of these organisms continue to be clarified.

Antigenic Types

The cell wall structure of group A streptococci is among the most studied of any bacteria. The cell wall is composed of repeating units of N-acetylglucosamine and N-acetylmuramic acid, the standard peptidoglycan. For decades, the definitive identification of streptococci has rested on the serologic reactivity of cell wall polysaccharide antigens originally delineated by Rebecca Lancefield. Eighteen group-specific antigens were established. The group A polysaccharide is a polymer of N-acetylglucosamine and rhamnose. Some group antigens are shared by more than one species; no Lancefield group antigen has been identified for *S pneumoniae* or for some other a- or g-streptococci. With advances in serologic methods, other streptococci have been shown to possess several established group antigens.

The cell wall also consists of several structural proteins. In group A streptococci, the R and T proteins may serve as epidemiologic markers, but the M proteins are clearly virulence factors associated with resistance to phagocytosis. More than 50 types of *S pyogenes* M proteins have been identified on the basis of antigenic specificity. Both the M proteins and lipoteichoic acid are supported externally to the cell wall on fimbriae, and the lipoteichoic acid, in particular, appears to mediate bacterial attachment to host epithelial cells. M protein, peptidoglycan, N-acetylglucosamine, and group-specific carbohydrate portions of the cell wall have antigenic epitopes similar in size and charge to those of mammalian muscle and connective tissue. Recently emerging strains of increased virulence are distinctly mucoid, rich in M protein and highly encapsulated.

The capsule of *S pyogenes* is composed of hyaluronic acid, which is chemically similar to that of host connective tissue and is therefore nonantigenic. In contrast, the antigenically reactive and chemically distinct capsular polysaccharide of *S pneumoniae* allows the single species to be separated into more than 80 serotypes. The antiphagocytic *S pneumoniae* capsule is the most clearly understood virulence factor of these organisms; type 3 *S pneumoniae*, which produces copious quantities of capsular material, are the most virulent. Unencapsulated *S pneumoniae* are avirulent. The polysaccharide capsule in *S agalactiae* allows differentiation into types Ia, Ib, Ic, II and III.

Finally, the cytoplasmic membrane of *S pyogenes* has antigens similar to those of human cardiac, skeletal, and smooth muscle, heart valve fibroblasts, and neuronal tissues, resulting in a molecular mimicry.

EXTRACELLULAR GROWTH PRODUCTS

The importance of the interaction of streptococcal products with mammalian blood and tissue components is becoming widely recognized. The soluble extracellular growth products or toxins of the streptococci, especially of *S pyogenes* (see Fig. 13-2), have been studied intensely. Streptolysin S is an oxygen-stable cytolysin; Streptolysin O is a reversibly oxygen-labile cytolysin. Both are leukotoxic, as is NADase. Hyaluronidase (spreading factor) can digest host connective tissue hyaluronic acid as well as the organism's own capsule. Streptokinases participate in fibrin lysis. Streptodornases A-D possess deoxyribonuclease activity; B and D possess ribonuclease activity as well. Protease activity similar to that in Staph aureus has been shown in strains causing soft tissue necrosis or toxic shock syndrome. This large repertoire of products may be important in the pathogenesis of *S pyogenes* by enhancing virulence; however, antibodies to these products appear not to protect the host even though they have diagnostic importance.

Three pyrogenic exotoxins of *S pyogenes* (SPEs) are recognized: types A, B, C. These toxins act as superantigens by a mechanism similar to those described for staphylococci, not requiring processing by antigen presenting cells. Rather, they stimulate T cells by binding class II MHC molecules directly and nonspecifically. With superantigens about 20% of T cells may be stimulated (vs 1/10,000 T cells stimulated by conventional antigens) resulting in massive detrimental cytokine release. When *S pyogenes* is lysogenized by certain bacteriophages, the SPEs A or C are produced; nonlysogenized strains are atoxic. SPE B is encoded by the bacterial chromosome. Re-emergence in the late 1980's of these exotoxin-producing strains has been associated with a toxic shock-like syndrome similar in pathogenesis and manifestation to staphylococcal toxic shock syndrome and other forms of invasive disease associated with severe tissue destruction. SPE's have also been identified from non group A streptococci (groups B, C, F. G) in association with the toxic shock-like syndrome.

Virulence factors in the other streptococcal species, including the enterococci, are less well identified. In group B streptococci, carbohydrate surface antigens associated with antiphagocytosis have been identified, as has neuraminidase, which may play a role in pathogenesis. Among the viridans streptococci, production of the exopolysaccharide (glycocalyx) is associated with the ability to adhere to the cardiac valves and to form vegetations on the valve leaflets.

Pathogenesis

Streptococcus pyogenes and Streptococcus pneumoniae

Streptococci vary widely in pathogenic potential. Despite the remarkable array of cell-associated and extracellular products previously described, no clear scheme of pathogenesis has been worked out. *S pneumoniae* and, to a lesser extent, *S pyogenes* are part of the normal human nasopharyngeal flora. Their numbers are usually limited by competition from the nasopharyngeal microbial ecosystem and by nonspecific host defense mechanisms, but failure of these mechanisms can result in disease. More often disease results from the acquisition of a new strain following alteration of the normal flora. *S pyogenes* causes inflammatory purulent lesions at the portal of entry, often the upper respiratory tract or the skin. Some strains of streptococci show a predilection for the respiratory tract; others, for the skin. Generally, streptococcal isolates from the pharynx and respiratory tract do not cause skin infections.

Invasion of other portions of the upper or lower respiratory tracts results in infections of the middle ear (otitis media), sinuses (sinusitis), or lungs (pneumonia). In addition, meningitis can occur by direct extension of infection from the middle ear or sinuses to the meninges or by way of bloodstream invasion from the pulmonary focus. Bacteremia can also result in infection of bones (osteomyelitis) or joints (arthritis).

S pyogenes (a group A streptococcus) is the leading cause of uncomplicated bacterial pharyngitis and tonsillitis.

Indeed, only group A streptococci are sought routinely in cases of pharyngitis, although groups B, C, and G are sometimes identified. *S pyogenes* infections can also result in sinusitis, otitis, mastoiditis, pneumonia with empyema, joint or bone infections, necrotizing fasciitis or myositis, and, more infrequently, in meningitis or endocarditis. *S pyogenes* infections of the skin can be superficial (impetigo) or deep (cellulitis). Although scarlet fever was formerly a severe complication of streptococcal infection, because of antibiotic therapy it is now little more than streptococcal pharyngitis accompanied by rash. Similarly, erysipelas, a form of cellulitis accompanied by fever and systemic toxicity, is less common today. There has, however, been an apparent recent increase in variety, severity and sequelae of *S pyogenes* infections. Because cases of streptococcal disease are not reported to national disease clearinghouses in the US, absolute numbers are not available. However, the recent resurgence of severe invasive infections has prompted descriptions of "flesh eating bacteria" in the news media. There has been no major change in susceptibility of *S pyogenes* to commonly used antibiotics but rather in the strain variations described above (antigenic types and extracellular growth products). However, a complete explanation for the decline and resurgence is not yet available.

The capsule of *S pneumoniae* renders it resistant to phagocytosis. The ability to evade this important host defense mechanism allows *S pneumoniae* to survive, multiply, and spread to various organs. The cell wall of S *pneumoniae* contains teichoic acid. The inflammatory response induced by Gram-positive cell walls differs from that induced by the endotoxin of Gram-negative organisms, but does include recruitment of polymorphonuclear neutrophils, changes in permeability and perfusion, cytokine release, and stimulation of platelet-activating factor. The role of other S pneumoniae moieties in virulence is less clear: protein A, pneumolysin, and peptide permeases. *S pneumoniae* is the leading cause of bacterial pneumonia beyond the neonatal period. Pleural effusion is the most common and empyema (pus in the pleural space) one of the most serious complications of *S pneumoniae*. This organism is also the most common cause of sinusitis, acute bacterial otitis media, and conjunctivitis beyond early childhood. Dissemination from a respiratory focus results in serious disease: outpatient bacteremia in children, meningitis, occasionally acute septic arthritis and bone infections in patients with sickle cell disease and, more rarely, peritonitis (especially in patients with nephrotic syndrome) or endocarditis.

Postinfectious Sequelae

Infection with *S pyogenes* (but not *S pneumoniae*) can give rise to serious nonsuppurative sequelae: acute rheumatic fever and acute glomerulonephritis. These sequelae begin 1-3 weeks after the acute illness, a latent period consistent with an immune-mediated rather than pathogen-disseminated etiology. Whether all *S pyogenes* strains are rheumatogenic is still controversial; however, clearly not all are nephritogenic. These differences in pathogenic potential are not yet understood.

Acute rheumatic fever is a sequela only of pharyngeal infections, but acute glomerulonephritis can follow infections of the pharynx or the skin. Although there is no adequate explanation for the precise pathogenesis of acute rheumatic fever or for its failure to occur after streptococcal pyoderma, an abnormal or enhanced immune response seems essential. Also, persistence of the organism, due perhaps in part to the greater avidity with which the organism adheres to host pharyngeal cells, is associated with an increased likelihood of rheumatic fever. Acute glomerulonephritis results from deposition of antigen-antibody-complement complexes on the basement membrane of kidney glomeruli.

The antigen may be streptococcal in origin or it may be a host tissue species with antigenic determinants similar to those of streptococcal antigen (cross-reactive epitopes for endocardium, sarcolemma, vascular smooth muscle). In the United States, the incidence of acute rheumatic fever had decreased dramatically. Although several areas reported a resurgence in cases in the late 1980's, subsequently, a slow, steady decline continued. Acute rheumatic fever can result in permanent damage to the heart valves. Less than 1% of sporadic streptococcal pharyngitis infections result in acute rheumatic fever; however, recurrences are common, and life-long antibiotic prophylaxis is recommended following a single case. The incidence of acute glomerulonephritis in the United States is more variable, perhaps due to cycling of nephritogenic strains, but appears to be decreasing; recurrences are uncommon, and prophylaxis following an initial attack is unnecessary.

OTHER STREPTOCOCCAL SPECIES

Lancefield Group Streptococci

Streptococcal groups B, C, and G initially were recognized as animal pathogens and as part of the normal human flora. Recently, the pathogenic potential for humans of some of these non-group-A streptococci has been clarified. Group B streptococci, a major cause of bovine mastitis, are a leading cause of neonatal septicemia and meningitis, accounting for a significant changing clinical spectrum of diseases in both pregnant women and their infants. Mortality rates in full-term infants range from 2-8% but in pre-term infants are approximately 30%. Early-onset neonatal disease (associated with sepsis, meningitis and pneumonia at 6d life) is thought to be transmitted vertically from the mother; late-onset (from 7d to 3 mos age) meningitis is acquired horizontally, in some instances as a nosocomial infection. Group B organisms also have been associated with pneumonia in elderly patients. They are part of the normal oral and vaginal flora and have also been isolated in adult urinary tract infection, chorioamnionitis and endometritis, skin and soft tissue infection, osteomyelitis, meningitis, bacteremia without focus, and endocarditis. Infection in patients with HIV can occur at any age.

Streptococci of groups C and G are associated with mild, as well as severe human disease. None of these groups has been implicated in acute rheumatic fever or acute glomerulonephritis. Group D streptococci are important etiologic agents of urinary tract infections and infections associated with biliary tract procedures, as well as cases of disseminated infection, bacteremia, and endocarditis. Streptococcus bovis bacteremia has been recognized more often in cases of bowel disease.

Group F streptococci are associated with abscess formation and purulent disease. Group R streptococci, well-documented causes of meningitis and septicemia in pigs, also pose a serious health hazard to workers in the pork industry.

Viridans Streptococci

The biochemically and antigenically diverse group of organisms classified as viridans streptococci, as well as other non-groupable streptococci of the oral and gastrointestinal cavities and urogenital tract, include important etiologic agents of bacterial endocarditis. Dental manipulation and dental disease with the associated transient bacteremia are the most common predisposing factors in bacterial endocarditis, especially if heart valves have been damaged by previous rheumatic fever or by congenital cyanotic heart disease. *S mutans* and *S sanguis* are odontopathogens responsible for the formation of dental plaque, the dense adhesive microbial mass that colonizes teeth and is linked to caries and other human oral disease. *S mutans* is the more cariogenic of the two species, and its virulence is directly related to its ability to synthesize glucan from fermentable carbohydrates as well as to modify glucan in promoting increased adhesiveness.

Anaerobes

Like their aerobic counterparts, anaerobic streptococci are part of the normal flora, particularly of the mouth and intestinal tract; they are also part of the normal flora of the upper respiratory and genital tracts and the skin. These anaerobic organisms are linked to a wide variety of serious mixed infections of the female genital tract as well as to brain, pulmonary, and abdominal abscesses.

Host Defenses

The streptococci are part of the endogenous microbial flora of the nasopharynx. Disease may result from circumvention of the normal specific or nonspecific host defense mechanisms. More often, both *S pyogenes* and *S pneumoniae* are exogenous secondary invaders following viral disease or disturbances in the normal bacterial flora.

In the normal host, nonspecific defense mechanisms prevent organisms from penetrating beyond the superficial epithelium of the upper respiratory tract. These mechanisms include mucociliary movement and the cough, sneeze and epiglottal reflexes. The host phagocytic system is a second line of defense against pathogens.

Organisms can be opsonized by activation of the classical or alternate complement pathway or by specific immunoglobulin binding.

The capsules of both *S pyogenes* and *S pneumoniae* allow the organisms to evade opsonization. The hyaluronic acid outer surface of *S pyogenes* is only weakly antigenic; however, protective immunity results from the development of type-specific antibody to the M protein of the fimbriae, which protrude from the cell wall through the capsular structure. This antibody, which follows respiratory and skin infections, is persistent. Presumably, IgA in the respiratory secretions and serum IgG are the important protective antibody classes. *S pyogenes* is rapidly killed following phagocytosis enhanced by specific antibody. Prompt, effective antibiotic treatment of streptococcal infections may preclude development of this persistent antibody. Evidence has shown that antibody to the erythrogenic toxin involved in scarlet fever is also long lasting.

This is the basis of the Dick test, an in vivo skin test, rarely used today, which measures host antitoxin. The capsular polysaccharides of *S pneumoniae* are highly antigenic and type-specific. Type-specific anticapsular antibodies to these T-independent antigens result in effective opsonization and host recovery. In untreated *S pneumoniae* infections, recovery clearly is due to opsonizing antibody. Even when adequate and appropriate antibiotic therapy is given, opsonizing antibody probably contributes significantly to recovery from pneumococcal disease. The normal host is somewhat resistant to *S pneumoniae* disease, but compromised hosts of several types are highly susceptible to serious infections: alcoholics, the semicomatose, very young, and very old individuals, patients who have undergone splenectomy, and patients with underlying diseases (specifically, chronic cardiac, pulmonary, or renal disease; sickle cell anemia; leukopenia; multiple myeloma; cirrhosis; and diabetes).

Cross-reactive antigens, especially of *S pyogenes* and various mammalian tissues, help explain the autoimmune responses that develop following some infections. The level of humoral response to infection with *S pyogenes* is greater in patients with rheumatic fever than in patients with uncomplicated pharyngitis. In addition, cell-mediated immunity may play a significant role in acute rheumatic fever.

Neonatal susceptibility to group B streptococci may result from immature neonatal phagocytic function, humoral immunity, or cell-mediated immunity, or from lack of passively acquired maternal antibody.

Evidence from the Rhesus monkey animal model in dental research shows that IgG may be a more important antibody class than IgA or IgM in protection against caries. Part of the reason may be that IgG is the antibody isotype most efficient at enhancing phagocytosis of *S mutans*. Cell-mediated immunity appears to participate in the protective host response against caries.

Epidemiology

The streptococci are widely distributed in nature and frequently form part of the normal human flora. Approximately 5-15% of humans carry *S pyogenes* or *S agalactiae* in the nasopharynx. *S pneumoniae* infects humans exclusively, and no reservoir is found in nature. The carrier rate of *S pneumoniae* in the normal human nasopharynx is 20-40%.

All ages, races, and sexes are susceptible to streptococcal disease. Because *S pneumoniae* is a particularly labile organism, sensitive to heat, cold, and drying, horizontal transmission requires close person-to-person contact. Infection is more likely at the extremes of life (<2 yr, > 65 yr), when host resistance is reduced, as described in the preceding section, or after the introduction of a more virulent strain. In the United States, pneumococcal disease is most prevalent during winter, coinciding with increased rates of acquisition but not necessarily of carriage. Alaskan natives have higher rates of invasive pneumococcal disease than do other American populations. The reason for this is unclear.

The incidence of respiratory disease attributed to *S pyogenes* peaks at about 6 years of age, and then again at 13 years of age, and is most common during late winter and early spring in temperate climates. Skin infections are more common among preschool-age children, and are most prevalent in late summer and early fall in temperate climates (when hot, humid weather prevails), and at all times in tropical climates. *S pyogenes* is spread by respiratory droplets or by contact with fomites used by the index individual, either patient or carrier. Skin infections often follow minor skin irritation, such as insect bites. There are occasional reports of streptococcal disease traced to rectal carriers, and of food-borne and vector-born outbreaks. In children, invasive disease with *S pyogenes* may follow varicella, or be associated with burns or malignancy; in adults with surgical or nonsurgical wounds or underlying medical problems, i.e., diabetes, cirrhosis, underlying peripheral vascular disease, or malignancy.

The world prevalence of the serious late sequelae of *S pyogenes* infections (acute rheumatic fever and acute glomerulonephritis) has shifted from temperate to tropical climates. In particular, acute rheumatic fever had ceased to be a major health concern in the US., despite no concomitant decline in group A streptococcal pharyngitis. These diseases previously affected persons with a low standard of living and limited access to medical care. Since 1985, there have been scattered outbreaks of acute rheumatic fever in some regions of the United States. Temporal and geographic clustering provides further evidence for "rheumatogenic" strains. Whether ethnic or racially determined factors affect this shift is not known.

Other streptococcal groups show striking epidemiologic features. An increasing prevalence of non-group-A as compared to group A streptococci in throats has been reported. Studies of the vaginal flora among women of child-bearing age show a *S agalactiae* carrier rate of 15-40%. Vertical transmission of the organisms to neonates of vaginally infected mothers ranges from 40-73%, but the incidence of neonates with disease (in contrast to colonized, healthy neonates) is low, 1-2%. *S suis* has been linked to meningitis among meat handlers. Isolation of *S milleri* or *S bovis* from the bloodstream should raise suspicion of immunosuppression or underlying disease visceral abscess formation or other bowel disease (including colon carcinoma).

In the United States, enterococci are the second most common nosocomial pathogenes associated with both endogenous colonization and patient-to-patient spread. A wide variety of infections results, especially urinary tract and surgical wound infections, with a marked propensity for antibiotic resistance. The widespread usage of newer cephalosporins, which have poor activity against enterococci, allows "break through" of enterococci as clinically significant isolates, the development of resistance in areas of heavy antibiotic use, and a selective advantage to these organisms.

DIAGNOSIS

Clinical

It is not usually possible to diagnose streptococcal pharyngitis or tonsillitis on clinical grounds alone. Accurate differentiation from viral pharyngitis is difficult even for the experienced clinician, and therefore the use of bacteriologic methods is essential. However, distinguishing acute streptococcal pharyngitis from the carrier state may be difficult.

When documented streptococcal pharyngitis is accompanied by an erythematous punctiform rash, the diagnosis of scarlet fever can be made. With streptococcal toxic shock syndrome, unlike staphylococcal toxic shock syndrome where the organism is elusive, there is often a focal infection or bacteremia. Criteria for diagnosis of streptococcal toxic shock syndrome include hypotension and shock, isolation of *S pyogenes*, as well as 2 or more of the following: ARDS, renal impairment, liver abnormality, coagulopathy, rash with desquamating soft tissue necrosis. The invasive, potentially fatal *S pyogenes* infections require early recognition, definitive diagnosis, and early aggressive treatment.

Rheumatic fever is a late sequela of pharyngitis and is marked by fever, polyarthritis, and carditis. A combination of clinical and laboratory criteria is used in the diagnosis of acute rheumatic fever. Since the original Jones criteria were published in 1944, these have been modified (1955), revised (1965, 1984) and updated (1992). The other late sequela, acute glomerulonephritis, is preceded by pharyngitis or pyoderma; is characterized by fever, blood in the urine (hematuria), and edema; and is sometimes accompanied by hypertension and elevated blood urea nitrogen (azotemia). Pneumococcal pneumonia is a life-threatening disease, often characterized by edema and rapid lobar consolidation.

Specimens For Direct Examination And Culture

S pyogenes is usually isolated from throat cultures. In cases of cellulitis or erysipelas thought to be caused by *S pyogenes*, aspirates obtained from the advancing edge of the lesion may be diagnostic. *S pneumoniae* is usually isolated from sputum or blood. Precise streptococcal identification is based on the Gram stain and on biochemical properties, as well as on serologic characteristics when group antigens are present.

Identification

Hemolysis should not be used as a stringent identification criterion. Bacitracin susceptibility is a widely used screening method for presumptive identification of *S pyogenes*; however, some *S pyogenes* are resistant to bacitracin (up to 10%) and some group C and G streptococci (about 3-5%) are susceptible to bacitracin. Some of the group B streptococci also may be bacitracin sensitive, but are presumptively identified by their properties of hippurate hydrolysis and CAMP positivity. *S pneumoniae* can be separated from other a-hemolytic streptococci on the basis of sensitivity to surfactants, such as bile or optochin (ethylhydrocupreine hydrochloride). These agents activate autolytic enzymes in the organisms that hydrolyze peptidoglycan.

In many instances, presumptive identification is not carried further. Serologic grouping has not been performed as often as it might be because of the lack of available methods and the practical constraints of time and cost; however, only serologic methods, provide definitive identification of the streptococci. The Lancefield capillary precipitation test is the classical serologic method. *S pneumoniae,* which lacks a demonstrable group antigen by the Lancefield test, is conventionally identified by the quellung or capsular swelling test that employs type-specific anticapsular antibody. Inspection of Gram-stained sputum remains a reliable predictor for initial antibiotic therapy in community-acquired pneumonia.

New methods for serogrouping that show sensitivity and specificity now are being explored. Organisms from throat swabs, incubated for only a few hours in broth, can be examined for the presence of *S pyogenes* using the direct fluorescent antibody or enzyme-linked immunosorbent technique. Additional rapid antigen detection systems for the group carbohydrate have become increasingly popular. However, the sensitivity (70-90%) of these currently available rapid tests for group A streptococcal carbohydrate does not allow exclusion of streptococcal pharyngitis without conventional throat culture (sensitivity of a single throat culture is 90-99%). A third generation assay, the optical immunoassay, is currently being evaluated. *S pneumoniae* can be identified rapidly by counterimmuno-electrophoresis, a modification of the gel precipitin method.

The coagglutination test, is a more sensitive modification of the conventional direct bacterial agglutination test. The Fc portion of group-specific antibody binds to the protein A of dead staphylococci, leaving the Fab portion free to react with specific streptococcal antigen. The attachment of antibody to other carrier particles in suspension (for example, latex) also is used. The fact that whole streptococcal cells can be used in recently developed methods circumvents the difficulties involved in extracting components that retain appropriate antigenic reactivity. These newer serogrouping methods should make it more practical to identify not only b-hemolytic isolates from the blood or normally sterile sites, but also a-and nonhemolytic strains. It has become increasingly important to identify more of these strains to avoid simply misclassifying them as contaminants. Such information will expand our understanding of the importance of non-group-A streptococci.

Serologic Titers

Antibodies to some of the extracellular growth products of the streptococci are not protective but can be used in diagnosis. The antistreptolysin O (ASO) titer which peak 2-4 wks after acute infection and anti-NADase titers (which peaks 6-8 weeks after acute infection) are more commonly elevated after pharyngeal infections than after skin infections. In contrast, antihyaluronidase is elevated after skin infections, and anti-DNase B rises after both pharyngeal and skin infections. Titers observed during late sequelae (acute rheumatic fever and acute glomerulonephritis) reflect the site of primary infection. Although it is not as well known as the ASO test, the anti-DNase B test appears superior because high-titer antibody is detected following skin and pharyngeal infections and during the late sequelae. Those titers should be interpreted in terms of the age of the patient and geographic locale. Although not used in diagnosis, bacteriocin production and phage typing of streptococci are employed in research and epidemiologic studies.

Control

Antibiotic Treatment

Penicillin remains the drug of choice for *S pyogenes*. It is safe, inexpensive, and of narrow spectrum, and there is no direct or indirect evidence of loss of efficacy. Prior to the 1990's, *S pneumoniae* was also uniformly sensitive to penicillin but a recent abrupt shift in the usefulness of penicillin has occurred. The group D enterococci are resistant to penicillins, including penicillinase-resistant penicillins such as methicillin, nafcillin, dicloxacillin, and oxacillin, and are becoming increasingly resistant to many other antibiotics. Group B streptococci are often resistant to tetracycline but remain sensitive to the clinically achievable blood levels of penicillin, even though they have penicillin minimal inhibitory concentrations (MIC) considerably higher than those of *S pyogenes*. Although the duration of penicillin therapy varies with the degree of invasiveness, streptococcal pharyngitis is generally adequately treated with 10 days of antibiotic therapy, and pneumococcal pneumonia with 7-14 days. If penicillin allergy occurs, an alternative drug for treating pharyngitis is erythromycin, although sporadic erythromycin and tetracycline resistance has been reported, leaving clindamycin or the newer macrolides as possible treatments.

The most important goal of therapy in acute streptococcal pharyngitis is still to prevent rheumatic fever. However, therapy also hastens clinical recovery, avoids suppurative complications and renders the patient non-infectious for others. In addition to antibiotics, the patient with *S pyogenes* myositis or necrotizing fasciitis requires surgical debridement. Lifelong prophylaxis against recurrences of rheumatic fever is achieved with long-acting penicillin or erythromycin. Sulfonamides will not eradicate the streptococcus and thus are not acceptable therapy for streptococcal pharyngitis, but sulfadiazine is effective for preventing recurrent attacks of rheumatic fever. Additional prophylactic coverage before some dental and surgical procedures is necessary in the presence of rheumatic heart disease or prosthetic heart valves. Although streptococcal pharyngitis is usually a benign, self-limited disease, therapy is important to prevent rheumatic fever. There is no convincing evidence that antibiotic therapy prevents glomerulonephritis. Disconcertingly, some patients in recent outbreaks of acute rheumatic fever do not give a history of preceding pharyngitis.

Methods of treating the asymptomatic pharyngeal carrier of *S pyogenes* remain controversial. Recent evidence suggests that up to 20% of children and young adults are carriers, the carrier state involves no risk to the carrier or to others, and it is frequently difficult to eradicate despite the exquisite sensitivity of the organism to penicillin in vitro. A similar failure of antibiotic therapy to eradicate nasopharyngeal carriage or to prevent reinfection with *S pneumoniae* also occurs.

Although antibiotic resistance in *S pneumoniae* is common in many parts of the world, in the United States such strains previously had a geographically limited focus. Recent widespread emergence of *S pneumoniae* resistant to penicillin and other antibiotics has become a microbial threat in the United States as well. Even cefotaxime and ceftriaxone resistance has been documented. Isolates must be carefully screened for susceptibility by oxacillin disc testing, with definitive MIC determination by the E test (A B Biodisk NA, Piscataway, NJ), a convenient and reliable method for detection of resistance to penicillin and extended spectrum cephalosporins.

It is inappropriate to universally treat of pregnant women who are carriers of group B streptococci, or their colonized neonates, for several reasons: the high carrier rate; cost; the associated high risk of penicillin hypersensitivity; the potential increase in infections with penicillin-resistant organisms; the difficulty in altering colonization of women (even when their sexual partners were also treated); and the low risk of neonatal disease. The controversy continues despite recent recommendations for universal screening of preg nant women and selective intrapartum chemoprophylaxis for screen-positive mothers with preterm labor, premature or prolonged rupture of membranes, fever in labor, multiple births or previous infants with group B streptococcal disease.

Clearly, penicillin has reduced the severe morbidity and mortality associated with *S pneumoniae*. The emergence of resistance has now forced re-evaluation of empiric therapy. Clinicians must report clusters of *S pneumoniae* infection and be aware of local patterns of resistance. Penicillin susceptible organisms show MICs 0.06 mg/ml, intermediate strains 0.1-1.0

mg/ml and high level resistant strains 2 mg/ml. For nonmeningeal infection by intermediate strains, parenteral penicillin at high dose can probably be used since the mechanism of resistance involves alteration in penicillin binding proteins (PBP) and saturation.

For meningeal infection with intermediate strains or any infection by high level resistant strains only ceftriaxone and cefotaxime retain sufficient activity. Resistance even to these extended spectrum cephalosporins was first reported for the US in 1991. At this writing only vancomycin remains uniformly effective but as discussed below, its use incurs potential for selection of vancomycin resistant enterococci (VRE) or risk of transferring vancomycin resistance from enterococci to *S pneumoniae*.

Currently, no single agent is reliably bactericidal against enterococci. Serious infections with group D enterococci often require a classic synergistic regime combining penicillin or ampicillin with an aminoglycoside, designed to weaken the cell wall with the ß-lactam and facilitate entry of the bacteriocidal aminoglycoside. Other ß-lactam drugs with good activity against enterococci include piperacillin and imipenem. An alternate drug of choice is vancomycin, but vancomycin-resistant strains of enterococci have been isolated. Nosocomial acquisition of these resistant organisms is of grave concern.

This antibiotic resistance among the streptococci/ enterococci is an increasing problem. Studies show that in vitro exchange of resistant DNA can occur in conjugation via plasmids and transposons, or in transduction with bacteriophages. The mechanisms involved in the in vivo genetic exchange are not clearly defined. Evidence is accumulating that other streptococci may be the important donors of resistance markers.

Transposon transfer is thought to be the most likely mechanism in *S pneumoniae*, although point mutations also occur. In the setting of heavy ß-lactam use, selective pressure is important in emergence of resistant strains. The first penicillin-resistant *S pneumoniae* were reported in 1967 in Australia and in 1974 in North America. In New Guinea, where the first penicillin-resistant strains were reported in 1971, one-third of *S pneumoniae* isolates from patients with severe pneumococcal disease were resistant by 1978.

In Hungary in 1992, 69% of *S pneumoniae* isolates were penicillin resistant. This resistance is not ß-lactamase mediated but due to alteration in PBP which results in decreased binding of penicillin by the organism, rendering the drug less effective and requiring higher concentrations for saturation. Some strains resistant to erythromycin or tetracycline also have been reported, as well as some multiply resistant strains. In South Africa, outbreaks of infection with strains of *S pneumoniae* resistant to ß-lactam antibiotics (penicillins and cephalosporins) as well as to tetracycline, chloramphenicol, erythromycin, streptomycin, clindamycin, sulfonamides, and rifampin were reported in 1977. Although antibiotic resistance among *S pneumoniae* was infrequent in the United States, a major shift occurred from 1988 to 1990, resulting in the present situation of 15-25% of *S pneumoniae* intermediately or completely resistant to penicillin. Communities with "low prevalence" have 5-10% resistance. Single or multiply resistant strains are transmitted person to person, especially in settings of frequent salivary exchange, antibiotic use and hand-to-hand transmission (as in day care centers) or of crowding (corrections facilities, homeless shelters, nursing homes, military training groups).

Control of the problem of emerging, antibiotic-resistant *S pneumoniae* is multifactorial: (1) surveillance for clusters of invasive disease, resistance and prevalent serotypes; (2) education of physicians and the public about antibiotic use (decrease unnecessary antibiotic use for obviously viral infections and decrease antibiotic prophylaxis for otitis by use of intermittent or expectant dosing or of non ß-lactam based prophylaxissulfa. Use topical treatment for impetigo, and short course therapies and narrow spectrum antibiotics); (3) adherence to infection control strategies in day care centers; (4) aggressive promotion of the current 23-valent *S pneumoniae* vaccine and support of efforts to design a new vaccine effective in those <2 years of age, analogous to the eminently successful Haemophilus influenzae type B vaccine where bacterial polysaccharide is conjugated to protein to elicit a T cell-dependent response.

Among the enterococci, resistance to a wide variety of common antibiotics has emerged, with some strains resistant to all currently available antibiotics. There is no clinically proven treatment effective against enterococci multiply resistant to lactams, aminoglycosides, and vancomycin. The emergence of such organisms poses a stunning management dilemma. Resistance among the enterococci can be either intrinsic or acquired (by de novo genetic mutation or acquisition of DNA from resistant organisms). Enterococcal resistance to lactams is also mediated by altered PBP as in pneumococci, allowing cell wall synthesis even in the presence of antibiotic, or much less commonly by ß-lactamase. Resistance to aminoglycosides is mediated by decreased uptake or aminoglycoside modifying enzymes, and to vancomycin by decreased cell wall affinity for glycopeptide antibiotics. Further research into the mechanisms of resistance and new class(es) of antibiotics is essential.

A final concern about emerging resistance among the enterococci is the potential for genetic transfer of resistance genes to more virulent pathogens: Staph aureus, *S pneumoniae* and even Gram-negative organisms. So significant is this threat of emerging enterococcal resistance that the Centers for Disease Control and Prevention has issued a document addressing national guidelines. These include recommendations for (1) education of physicians and the public about the impact of vancomycin resistant enterococci (VRE), (2) vigilant surveillance for and detection of VRE, (3) strict enforcement of infection control strategies in hospitals, and (4) prudent vancomycin use or monotherapeutic use of extended spectrum cephalosporins. In a recent study of vancomycin use in US hospitals, use was about equally divided for treatment of a specific isolate, for prophylaxis, and for empiric coverage. The recommendations discourage vancomycin use for routine surgical prophylaxis, empiric prophylaxis in the patient with febrile neutropenia, the low birth weight infant or patients with vascular or peritoneal catheters, treatment of a single blood culture positive for coagulase-negative staphylococci, primary treatment of antibiotic-associated colitis, attempted eradication of colonization by methicillin-resistant Staph aureus (MRSA), or selective decontamination of the gastrointestinal tract.

Vaccination

As chemotherapeutic management becomes more difficult because of the threat of resistance, prevention becomes more important. With the introduction of antibiotics, previously successful pneumococcal vaccines fell into disuse. However, although prompt treatment with antibiotics has reduced the serious consequences of *S pneumoniae* infections (pre-antibiotic mortality rate of 30%), the disease incidence remains unchanged, and attention has been redirected to vaccines for *S pneumoniae* as well as for other streptococci. Pneumococcal vaccines (containing the pneumococcal polysaccharides of the most prevalent serotypes) have been licensed in several countries, including the United States. Initial use shows them to be useful and safe, but they remain under-utilized.

The spectre of multidrug resistant *S pneumoniae* may provide a new incentive for their use. In 1983, the United States Food and Drug Administration licensed a vaccine containing 23 serotypes, representing coverage against nearly 89% of the pneumococcal isolates submitted to the CDC in the 1987-1988 National Surveillance Study. The population target of pneumococcal vaccines includes those at high risk for serious pneumococcal disease: the elderly (65 and older) and children (2 years of age and older) with sickle cell anemia, with an immunocompromised state (lymphoma, asplenia, myeloma, acquired immunodeficiency syndrome), with nephrotic syndrome, or with chronic cardiopulmonary disease. Vaccines for the other streptococci remain experimental.

Vaccine production for the streptococci presents several formidable problems. For both *S pyogenes* and *S pneumoniae*, a large number of serotypes must be included in effective vaccines since successful selection of a common epitope remains elusive. Continuing surveillance to determine prevalent serotypes is necessary to insure that the vaccine formulations remain appropriate. For *S pyogenes*, it is critical to determine rheumatogenic and nephritogenic strains to limit the required multivalency of the vaccines. Alternatively a newly described conserved portion of M protein is a distant goal. Toxicity has been associated with M protein preparations, but lack of immunogenicity in highly purified preparations of antigens is still a problem. With streptococcal vaccines, the potential risk of antigenic cross-reactivity with cardiac tissue and an associated increased risk of acute rheumatic fever must be appreciated.

In group B neonatal disease chemoprophylaxis does not appear as practical as vaccine control. Passive immunity in group B streptococcal neonatal infection appears protective. Polyvalent hyperimmune gamma globulin and human monoclonal IgM antibody which reacts with multiple serotypes are undergoing efficacy studies. Active immunization of pregnant women with undegraded sialic acid-containing polysaccharide group B antigens is another important aspect of control.

The streptococci are ubiquitous, and their significance in medicine is remarkable. Exciting advances are being made in diagnosis and in understanding the mechanisms of pathogenesis, as well as in control of these well-known organisms. Problems with antibiotic resistance must preclude complacency in dealing with these common pathogens.

12

Neisseria, Moraxella, Kingella and Eikenella

INTRODUCTION

The family Neisseriaceae comprises the genera Neisseria, Moraxella, Kingella, and Acinetobacter. The only significant human pathogens are N gonorrhoeae, the agent of gonorrhea, and N. meningitidis, an agent of acute bacterial meningitidis. N gonorrhoeae infections have a high prevalence and low mortality, whereas N meningitidis infections have a low prevalence and high mortality.

Gonococcal infections are acquired by sexual contact and usually affect mucous membranes of the urethra in men and the endocervix in women, although the infection may disseminate to a variety of tissues. The pathogenic mechanism involves the attachment of the gonococci to nonciliated epithelial cells via pili (fimbriae) and the production of cytotoxic factors (endotoxin). Similarly, the lipopolysaccharide of meningococci is highly toxic, but an additional virulence factor is the antiphagocytic capsule. Both pathogens produce proteases that cleave and inactivate human immunoglobulin A1 (IgA1), a major mucosal immunoglobulin of humans. Many normal individuals harbor meningococci, whereas gonococci are present only if sexual contact with an infected person has occurred. Epidemics of meningococcal meningitis occur sporadically. Gonococcal infections occur frequently and affect large numbers of sexually active people. Other species in this genus are primarily parasites on mucosal surfaces of humans and other animals. Human disease caused by these organisms usually is associated with opportunistic infections in compromised patients.

Nesseria gonorrhoeae

Clinical Manifestations

Gonorrheal infection is generally limited to superficial mucosal surfaces lined with columnar epithelium. The areas most frequently involved are the cervix, urethra, rectum, pharynx, and conjunctiva. Squamous epithelium, which lines the adult vagina, is not susceptible to infection by the gonococcus. However, the prepubertal vaginal epithelium, which has not been keratinized under the influence of estrogen, may be infected. Hence, gonorrhea in young girls may present as vulvovaginitis. Mucosal infections are usually characterized by a marked local neutrophilic response (purulent discharge).

The most common symptom of uncomplicated gonorrhea is a discharge that may range from a scanty, clear, or cloudy fluid to one that is copious and purulent. Dysuria is often present. Men with asymptomatic urethritis are an important reservoir for transmission. In addition, such men and those who ignore their symptoms are at increased risk for developing complications.

Endocervical infection is the most common form of uncomplicated gonorrhea in women. Such infections are usually characterized by vaginal discharge and sometimes by dysuria (because of coexistent urethritis). The cervical os may be erythematous and friable, with a purulent exudate. About 50 percent of women with cervical infections are asymptomatic. Local complications include abscesses in Bartholin's and Skene's glands.

Rectal infections with N gonorrhoeae occur in about one-third of women with cervical infection. They most often result from autoinoculation with cervical discharge and are rarely symptomatic. Rectal infections in homosexual men usually result from anal intercourse and are more often symptomatic. The symptoms and signs of gonococcal proctitis range from mild burning on defecation to itching to severe tenesmus and from mucopurulent discharge to frank blood in the stools. Pharyngeal infections are diagnosed most often in women and homosexual men with a history of fellatio. Such infections may be a focal source of gonococcemia.

Ocular infections can have serious consequences (corneal scarring or perforation); prompt diagnosis and treatment are therefore important. Ocular infections (ophthalmia neonatorum) occur most commonly in newborns who are exposed to infected secretions in the birth canal. Keratoconjunctivitis is occasionally seen in adults as a result of autoinoculation.

Disseminated gonococcal infections result from gonococcal bacteremia. Asymptomatic infections of the pharynx, urethra, or cervix often serve as focal sources for bacteremia. The most common form of disseminated gonococcal infection is the dermatitis-arthritis syndrome. It is characterized by fever, chills, skin lesions, and arthralgias (usually involving the hands, feet, and elbows), which are due to periarticular inflammation of the tendon sheaths. Occasionally, a patient develops a septic joint with effusion. Skin lesions may be macular, pustular, centrally necrotic, or hemorrhagic. Rarely, disseminated gonococcal infection causes endocarditis or meningitis. Gonococci may ascend from the endocervical canal through the endometrium to the fallopian tubes and ultimately to the pelvic peritoneum, resulting in endometritis, salpingitis, and finally, peritonitis. Women usually present with pelvic and abdominal pain, fever, chills, and cervical motion tenderness.

This complex of signs and symptoms is referred to as pelvic inflammatory disease (PID). This disease may also be caused by other sexually transmitted organisms (e.g., Chlamydia trachomatis) as well as by non-sexually transmitted bacteria that are part of the normal vaginal flora. Complications of pelvic inflammatory disease include tubo-ovarian abscesses, pelvic peritonitis, or Fitz-Hugh and Curtis syndrome, which is an inflammation of Glisson's capsule of the liver. As many as 15 percent of women with uncomplicated cervical infections may develop pelvic inflammatory disease. The disease may have serious consequences, including an increased probability of infertility and ectopic pregnancy.

Structure

Neisseria species are Gram-negative cocci, 0.6 to 1.0 μm in diameter. The organisms are usually seen in pairs with the adjacent sides flattened. Pili, hairlike filamentous appendages extend several micrometers from the cell surface and have a role in adherence. The outer membrane is composed of proteins, phospholipids, and lipopolysaccharide (LPS). Features that distinguish gonococcal LPS from enteric LPS are the highly branched basal oligosaccharide structure and the absence of repeating O-antigen subunits. For these reasons gonococcal LPS, as well as that of other mucosal pathogens, is referred to as lipooligosaccharide (LOS). Gonococci characteristically release outer membrane fragments (blebs) during growth. These blebs contain LOS and may have a role in pathogenesis.

CLASSIFICATION AND ANTIGENIC TYPES

The gonococcus is an obligate human pathogen. It is one of two Neisseria species that cause significant human infections. The genus also includes several nonpathogenic species, which may be part of the normal flora and therefore can be confused with N gonorrhoeae. Gonococcal strains can be characterized according to their nutritional requirements (auxotyping). A panel of monoclonal antibodies specific for epitopes on protein I have also been used to type strains. Strains exhibiting specific reaction patterns are termed serovars. A combined auxotype-serovar classification provides greater resolution among gonococcal isolates and is useful in epidemiologic investigations.

Pathogenesis

Our knowledge of the molecular basis of gonococcal pathogenesis is incomplete. Attachment of gonococci to mucosal cells is mediated in part by pili, although nonspecific factors such as surface charge and hydrophobicity may be important. Pili undergo both phase and antigenic variation. Opa proteins (protein II), which are located in the outer membrane, are also involved in attachment to host cells. Gonococci attach only to microvilli of nonciliated columnar epithelial cells; attachment to ciliated cells is not observed.

Much of our knowledge of gonococcal invasion comes from studies with tissue culture cells and human fallopian tube organ culture. After gonococci attach to the nonciliated epithelial cells of the fallopian tube, they are surrounded by the microvilli, which draw them to the surface of the mucosal cell. The gonococci appear to enter the epithelial cells by a process called parasite-directed endocytosis. This process seems to be initiated by microbial factors because it does not occur unless the gonococci are viable and because it involves host cells that are not normally phagocytic. An unidentified factor in serum enhances engulfment of gonococci. The process is inhibited by drugs that block the actions of the microtubule (demecolcine) and microfilament (cytochalasin B) systems. During endocytosis the membrane of the mucosal cell retracts, pinching off a membrane-bound vacuole that contains gonococci; this vacuole is rapidly transported to the base of the cell, where gonococci are released by exocytosis into the subepithelial tissue. Gonococci are not destroyed within the phagocytic vacuole; it is not clear whether they replicate in the vacuoles.

The major porin protein of the gonococcal outer membrane, Por (protein I), has been proposed as a candidate invasin (a substance that helps mediate invasion into a host cell). The insertion of Por into neutrophils treated with the chemotactic peptide, fMLP and leukotriene B4, inhibits degranulation but not the generation of the superoxide anion. The significance of these observations with respect to the pathogenesis of gonorrhea remains to be determined. Each gonococcal strain expresses only one type of Por; however, the Por of different strains may exhibit antigenic differences.

Gonococci can produce one or several outer membrane proteins called Opa proteins (proteins II). These proteins are subject to phase variation and are usually found on cells from colonies possessing an opaque phenotype (O+). At any one time, a gonococcus may express zero, one, or several different Opa proteins, though each strain has 10 or more genes for different Opas. Trypsin-like proteases present in cervical mucus may help select for protease-resistant transparent (O-) colony phenotypes. O+ colony phenotypes (protease sensitive) predominate in cultures taken during the middle portion of the menstrual cycle. Cervical proteases increase during the second half of the cycle, resulting in an increase in the O- phenotype. The O- colony types can be isolated from tubal as well as endocervical cultures; O+ colony phenotypes have been isolated more often from endocervical cultures than from tubal cultures.

Rmp (Protein III) is an outer membrane protein found in all strains of N gonorrhoeae. It does not undergo phase variation and is found in a complex with Por and LOS. It shares partial homology with the Omp A protein of Escherichia coli. Antibodies to Rmp, induced either by a neisserial infection or by colonization with E coli, block bactericidal antibodies directed against Por and LOS. Rmp antibodies may facilitate infection with N gonorrhoeae.

LOS has a profound effect on the virulence and pathogenesis of N gonorrhoeae. Gonococci can express several antigenic types of LOS and can alter the type of LOS they express by an as yet unknown mechanism. Gonococcal LOS produces mucosal damage in fallopian tube organ cultures and brings about the release of enzymes, such as proteases and phospholipases, that may be important in pathogenesis. More recent evidence suggests that gonococcal LOS stimulates the production of tumor necrosis factor (TNF) in fallopian tube organ cultures; inhibition of tumor necrosis factor with specific antiserum prevents tissue damage. Thus, gonococcal LOS appears to have an indirect role in mediating tissue damage. Gonococcal LOS is also involved in the resistance of N gonorrhoeae to the bactericidal activity of normal human serum. Oligosaccharides containing epitopes defined by specific monoclonal antibodies are associated with a serum-resistant phenotype.

Gonococci can utilize host-derived cytidine monophospho-N-acetylneuraminic acid (CMP-NANA) in vivo to sialylate the oligosaccharide component of its LOS, converting a serum-sensitive organism to a serum-resistant one. When such organisms are grown in vitro without CMP-NANA, their resistance to killing by normal human serum is rapidly lost. Organisms with non-sialylated LOS are more invasive than those with sialylated LOS. There is antigenic similarity between neisserial LOS and antigens present on human erthyrocytes. This similarity to self may preclude an effective immune response to these LPS antigens.

Gonococci are highly autolytic and release peptidoglycan fragments during growth. These fragments, released by bacterial and/or host peptidoglycan hydrolases, are toxic for fallopian tube mucosa and may contribute to the intense inflammatory reactions characteristic of gonococcal disease.

N gonorrhoeae is highly efficient at utilizing transferrin-bound iron for in vitro growth; many strains can also utilize lactoferrin-bound iron. Gonococci (and meningococci) bind only human transferrin and lactoferrin. This specificity is thought to be the reason these organisms are exclusively human pathogens. Nevertheless, the role of transferrin- and lactoferrin-bound iron in in vivo growth is unknown. Gonococci express several new proteins when grown under iron-restricted conditions similar to the conditions occurring in the host. Some of these proteins function as receptors for transferrin, lactoferrin, heme, and hemoglobin; others function in the transport of iron into the cell. Gonococci cannot grow anaerobically unless low concentrations of the alternative electron acceptor nitrite are present.

Under these conditions they produce novel proteins. These proteins are apparently produced during an infection because antibodies against them are present in the serum specimens of patients with uncomplicated gonorrhea, disseminated gonococcal infection, or pelvic inflammatory disease. These data suggest that some gonococci in the host are growing under anaerobic conditions. Further studies will determine the relevance of these proteins to pathogenesis.

Strains of N gonorrhoeae (and N meningitidis) produce two distinct extracellular IgA1 proteases, which cleave the heavy chain of human immunoglobulin A1 (IgA1) at different points within the hinge region. Type 1 protease cleaves a prolyl-seryl peptide bond and type 2 protease cleaves a prolyl-threonyl bond in the hinge region of the heavy chain. This region is missing in human IgA2, and so this isotype is not susceptible to cleavage. Each gonococcal or meningococcal isolate elaborates only one of these two enzymes. Split products of IgA1 have been found in the genital secretions of women with gonorrhea, suggesting that the gonococcal IgA1 protease is present and active during genital infection. Fab fragments of IgA1 may bind to the gonococcal cell surface and block the Fc-mediated functions of intact immunoglobulins.

Host Defenses

Not everyone exposed to N gonorrhoeae acquires the disease. This may be due to variations in the size or virulence of the inoculum, to nonspecific resistance, or to specific immunity. A 50 percent infective dose (ID50) of about 1,000 organisms has been established, based on the experimental urethral inoculation of male volunteers. There is no reliable ID50 for women, although it is assumed to be similar.

Nonspecific factors have been implicated in natural resistance to gonococcal infection. In women, changes in the genital pH and hormones may increase resistance to infection at certain times of the menstrual cycle. Urinary solutes exhibit bactericidal and bacteriostatic activity of N gonorrhoeae. Factors in urine that seem to be important are pH, osmolarity, and the concentration of urea. The variability in the susceptibility of gonococcal strains to the bactericidal and bacteriostatic properties of urine is thought to be one of the reasons some men do not develop a gonococcal infection when exposed.

Most uninfected individuals have serum antibodies that react with gonococcal antigens. These antibodies probably result from colonization or infection with various Gram-negative bacteria that possess cross-reactive antigens. These "natural" antibodies differ, both qualitatively and quantitatively, from person to person, but may be important in an individual's natural resistance or susceptibility to infection.

Infection with N gonorrhoeae stimulates both mucosal and systemic antibodies to a variety of gonococcal antigens. Mucosal antibodies are primarily IgA and IgG. In genital secretions, antibodies have been identified that react with Por, Opa and LOS, and some of the iron-regulated proteins. Vaccine trials have suggested that antipilus antibodies inhibit the pilus-mediated attachment of the homologous gonococcal strain. Complement is present in endocervical secretions, but in a much lower concentration than in blood. However, there is little evidence to support a role for a complement-mediated bactericidal defense mechanism on the genital mucosa. In general, the IgA response is brief and declines rapidly after treatment; IgG levels decline more slowly.

More information is available about the function of systemic humoral immune mechanisms in gonococcal infection. Gonococcal antigens such as pili, Por, Opa, Rmp, and LOS elicit a serum antibody response during an infection. Antipilus antibody levels tend to be higher in women than in men and are related to the number of previous infections. The predominant IgG subclass that reacts with a variety of gonococcal antigens is IgG3, followed by IgG1 and IgG4. IgG2 is minimal, suggesting that polysaccharides are not important in the immune response to gonococcal infection. Anti-Por antibodies may be bactericidal for the gonococcus. IgG that reacts with Rmp blocks the bactericidal activity of antibodies directed against Por and LOS. Genital infection with N gonorrhoeae stimulates a serum antibody response against the LOS of the infecting strain. Disseminated gonococcal infection results in higher levels of anti-LOS antibody than do genital infections.

Strains that cause uncomplicated genital infections usually are killed by normal human serum and are termed serum sensitive. This bactericidal activity is mediated by IgM and IgG that recognize sites on the LOS. Strains that cause disseminated infections are not killed by most normal human serum and are referred to as serum resistant. Resistance is mediated, in part, by IgA that blocks the IgG-mediated bactericidal activity of the serum. Serum specimens from convalescent patients with disseminating infections contain bactericidal IgG to the LOS of the infecting strain.

Individuals with inherited complement deficiencies have a markedly increased risk of acquiring systemic neisserial infections and are subject to recurring episodes of systemic gonococcal and meningococcal infections, indicating that the complement system is important in host defense.Gonococci activate complement by both the classic and alternative pathways.

Complement activation by gonococci leads to the formation of the C5b-9 complex (membrane attack complex) on the outer membrane. In normal human serum, similar numbers of C5b-9 complexes are deposited on serum-sensitive and serum-resistant organisms, but the membrane attack complex is not functional on serum-resistant organisms.

Other complement-mediated functions, such as opsonophagocytosis and chemotaxis, are more efficient with serum-sensitive than with serum-resistant gonococci. This may be a significant factor in the pathogenesis of disseminated gonococcal infection and probably contributes to the relative lack of genital symptoms observed with this disease.

Normal human serum contains opsonic anti-Por IgG. Antibodies to various surface-exposed antigens are also present in cervical and urethral secretions of patients with gonorrhea and probably contribute to the opsonophagocytosis of the organism. Opa is important in gonococcus-neutrophil interactions.

Gonococci expressing certain Opas interact with neutrophils in the absence of antibodies. Once phagocytosed, gonococci are killed by both oxygen-dependent and oxygen-independent mechanisms. The survival of gonococci within neutrophils has been the subject of considerable controversy, with no clear-cut answer yet available. The opsonization and phagocytosis of gonococci are comparatively more important in mucosal infections than in protection from systemic gonococcal (and meningococcal) infections.

Epidemiology

The only natural host for N gonorrhoeae is the human. Gonorrhea has all but disappeared in Scandinavia and several other European countries. In the United States, gonorrhea remains the most frequently reported infectious disease.

Between 1977 and 1993, the number of reported cases decreased 56 percent, from 1 million to 439,673 cases per year. The Centers for Disease Control (CDC) estimates that there are two unreported cases for every reported case of gonorrhea.

Gonorrhea is transmitted almost exclusively by sexual contact. The highest rates occur in women between the ages of 15 and 19 years and in men 20 and 24 years of age. Persons who have multiple sex partners are at highest risk. Rates of gonorrhea are higher in males and in minority and inner-city populations.

Gonorrhea is usually contracted from a sex partner who is either asymptomatic or has only minimal symptoms. It is estimated that the efficiency of transmission after one exposure is about 35 percent from an infected woman to an uninfected man and 50 to 60 percent from an infected man to an uninfected woman.

More than 90 percent of men with urethral gonorrhea will develop symptoms within 5 days; fewer than 50 percent of women with anogenital gonorrhea will do so. Women with asymptomatic infections are at higher risk of developing pelvic inflammatory disease and disseminated gonococcal infection.

Diagnosis

Gonococcal infection produces several common clinical syndromes that have multiple causes or that mimic other conditions. Laboratory tests are often required to differentiate among the etiologic agents causing urethritis or cervicitis.

The etiologic diagnosis of salpingitis and pelvic peritonitis is quite difficult because mixed infections are common and laparoscopy is required to obtain appropriate cultures. Gonococcal perihepatitis may mimic acute cholecystitis.

All of the above syndromes are also caused by C trachomatis, a sexually transmitted bacterium that causes more infections in the United States than N gonorrhoeae. The gonococcal arthritis-dermatitis syndrome, must be, differentiated from meningococcemia and Reiter syndrome, in particular, and from other causes of septic arthritis.

Customarily, the laboratory diagnosis of gonorrhea is made presumptively and then confirmed; the latter process involves identifying characteristics that distinguish N gonorrhoeae from other Neisseria spp. that may be present in the specimen. Nonpathogenic Neisseria are normal inhabitants of the oropharynx and nasopharynx and occasionally are isolated from other sites infected by N gonorrhoeae.

A presumptive diagnosis of gonorrhea may be made from Gram-stained smears of urethral, cervical, and rectal specimens if Gram-negative diplococci are observed within leukocytes; it is equivocal if only extracellular Gram-negative diplococci are seen and negative if no Gram-negative diplococci are seen. Gram stain diagnosis has a sensitivity and specificity of >95 percent in men with symptomatic urethritis. The specificity of Gram stain diagnosis in women is also high if the cervix is wiped clean to remove cervical secretions before collecting the specimen; however, the sensitivity is only about 50 percent. The sensitivity and specificity of the Gram stain for rectal specimens are lower than with cervical specimens.

Specimens for the laboratory diagnosis of gonorrhea should be collected before treating the patient. Ideally, specimens should be inoculated onto appropriate media and incubated immediately after collection at 35 to 36.5°C in a CO_2-enriched atmosphere, which can be obtained by using a candle extinction jar or a CO_2 incubator.

Urethral specimens are normally obtained from heterosexual men; urethral, rectal, and pharyngeal specimens are normally obtained from homosexual men; and cervical and rectal specimens are normally obtained from women. Specimens are collected with cotton, polyester, or calcium alginate swabs. When appropriate, specimens may also be obtained from the urethra and from Bartholin's and Skene's glands of infected women. Blood cultures should be performed for patients with suspected disseminated infection. Synovial fluid cultures should be performed for patients with septic arthritis.

Urethral, cervical, and pharyngeal specimens are inoculated onto selective medium such as modified Thayer-

Martin, Martin-Lewis, or NYC medium. These are complex media that contain antimicrobial and antifungal agents to inhibit the growth of unwanted organisms.

Rectal specimens should be inoculated onto modified Thayer-Martin medium which contains trimethoprim lactate to inhibit the growth and swarming of Proteus species. Specimens collected from normally sterile sites such as blood, synovial fluid, and conjunctivae may be inoculated onto a nonselective medium such as chocolate agar.

The combination of oxidase-positive colonies and Gram-negative diplococci provides a presumptive identification of N gonorrhoeae. Fluorescent-antibody staining, coagglutination, specific biochemical tests, and DNA probes may be used for confirmation. DNA probes have also been used to detect gonococci in urethral and cervical specimens. A commercial test based on this approach is available. Serologic tests for uncomplicated gonorrhea have not proved satisfactory.

Control

There is no effective vaccine to prevent gonorrhea. Candidate vaccines consisting of pilus protein or Por are of little benefit. The development of an effective vaccine has been hampered by the lack of a suitable animal model and the fact that an effective immune response has never been demonstrated. Condoms are effective in preventing the transmission of gonorrhea.

Contact tracing to identify source contacts (i.e., those who infected the index patient) has been useful in identifying asymptomatic individuals or those with ignored symptoms. Contact tracing has also been used to identify contacts who were exposed to the index patient and who may have become infected.

The evolution of antimicrobial resistance in N gonorrhoeae may ultimately affect the control of gonorrhea. Strains with multiple chromosomal resistance to penicillin, tetracycline, erythromycin, and cefoxitin have been identified in the United States and most other parts of the world. Sporadic high-level resistance to spectinomycin and fluoroquinolones have been reported.

Penicillinase-producing strains of N gonorrhoeae were first described in 1976. Five related ß-lactamase plasmids of different sizes have been identified in these strains. The strains cause more than one-half of all gonococcal infections in parts of Africa and Asia.

Their prevalence has increased dramatically in the United States since 1984 and has affected nearly every major metropolitan area. Plasmid-mediated high-level resistance of N gonorrhoeae to tetracycline was first described in 1986 and has now been reported in most parts of the world. This resistance is due to the presence of the streptococcal tetM determinant on a gonococcal conjugative plasmid.

The current CDC Treatment Guidelines recommend treatment of all gonococcal infections with antibiotic regimens effective against resistant strains. The recommended antimicrobial agents are ceftriaxone, cefixime, ciprofloxacin, or oflaxacin. Since a significant proportion of patients with gonorrhea are also infected with C trachomatis, doxycycline or erythromycin has been added to treat this concomitant infection.

NEISSERIA MENINGITIDIS

Clinical Presentation

N meningitidis infection results from the bloodborne dissemination (meningococcemia) of the meningococcus, usually following an asymptomatic or mildly symptomatic nasopharyngeal carrier state or a mild rhinopharyngitis. The mildest form is a transient bacteremic illness characterized by a fever and malaise; symptoms resolve spontaneously in 1 to 2 days. Acute meningococcemia is more serious and is often complicated by meningitis.

The manifestations of meningococcal meningitis are similar to acute bacterial meningitis caused by organisms such as Streptococcus pneumoniae, Haemophilus influenzae, and E coli. The manifestations result from both infection and increased intracranial pressure. Chills, fever, malaise, and headache are the usual manifestations of infection; headache, vomiting, and rarely, papilledema may result from increased intracranial pressure. Signs of meningeal inflammation are also present. The onset of meningococcal meningitis may be abrupt or insidious.

Infants with meningococcal meningitis rarely display signs of meningeal irritation. Irritability and refusal to take food are typical; vomiting occurs early in the disease and may lead to dehydration. Fever is typically absent in children younger than 2 months of age.

Hypothermia is more common in neonates. As the disease progresses, apneic episodes, seizures, disturbances in motor tone, and coma may develop. In older children and adults, specific symptoms and signs are usually present, with fever and altered mental status the most consistent findings. Headache is an early, prominent complaint and is usually very severe. Nausea, vomiting, and photophobia are also common symptoms.

Neurologic signs are common; approximately one-third of patients have convulsions or coma when first seen by a physician. Signs of meningeal irritation such as cervical rigidity (Brudzinski sign), thoracolumbar rigidity, hamstring spasm (Kernig sign), and exaggerated reflexes are common.

Petechiae (minute hemorrhagic spots in the skin) or purpura (hemorrhages into the skin) occurs from the first to the third day of illness in 30 to 60 percent of patients with meningococcal disease, with or without meningitis. The lesions may be more prominent in areas of the skin subjected to pressure, such as the axillary folds, the belt line, or the back.

Fulminant meningococcemia (Waterhouse-Friderichsen syndrome) occurs in 5 to 15 percent of patients with meningococcal disease and has a high mortality rate. It begins abruptly with sudden high fever, chills, myalgias, weakness, nausea, vomiting, and headache.

Apprehension, restlessness, and frequently, delirium occur within the next few hours. Widespread purpuric and ecchymotic skin lesions appear suddenly. Typically, no signs of meningitis are present. Pulmonary insufficiency develops within a few hours, and many patients die within 24 hours of being hospitalized despite appropriate antibiotic therapy and intensive care.

Structure

The only distinguishing structural feature between N meningitidis and N gonorrhoeae is the presence of a polysaccharide capsule in the former. The capsule is antiphagocytic and is an important virulence factor.

Bibliography

Du Pasquier, L. and Litman, G. W.: *Origin and Evolution of the Vertebrate Immune System*, Berlin, Springer, 2000.

Flajnik, M. F., Miller, K. and Du Pasquier, L.: *Fundamental Immunology*, Philadelphia, Lippincott Williams & Wilkins, 2003.

Hoffman, J. A., Janeway, C. A. and Natori, S.: *Phylogenetic Perspectives in Immunity: The Insect Host Defense*, Austin, R. G. Landes Company, 1994.

Kelsoe, G. and Schulze, D. H.: *Evolution and Vertebrate Immunity: The Antigen-Receptor and MHC Gene Families*, Austin, University of Texas Press, 1987.

Klein, J.: *Natural History of the Major Histocompatibility Complex*, New York, John Wiley & Sons, 1986.

Manning, M. J.: *Phylogeny of Immunological Memory*, Developments in Immunology, Amsterdam, Elsevier/North-Holland Biomedical Press, 1980.

Marchalonis, J. J.: *Immunity in Evolution*. Cambridge, Mass., Harvard University Press, 1976.

Nonaka, M.: *Origin and Evolution of the Vertebrate Immune System*, Berlin, Springer, 2000.

Síma, P. and Vetvicka, V.: *Evolution of Immune Reactions*, Boca Raton, CRC Press, 1990.

Vetvicka, V. and Síma, P.: *Evolutionary Mechanisms of Defense Reactions*, Basel, Birkhauser Verlag, 1998.

Vetvicka, V., Síma, P., Cooper, E. L., Bilej, M. and Roch, P.: *Immunology of Annelids*, Boca Raton, CRC Press, 1994.

Warr, G. W. and Cohen, N.: *Phylogenesis of Immune Functions*, Boca Raton, CRC Press, 1991.

Index

□□□